Weird and wonderful…

Arum maculatum

When ripe for pollination, lords and ladies (Arum maculatum) gives off a powerful odour to attract insects. Midges crawl on the female flowers at the base of the tube (the males are above). The central hairs allow midges to enter but not to leave until the pollen is shed on them.

Green Inheritance

PALM BEACH COUNTY
LIBRARY SYSTEM
3650 SUMMIT BLVD
WEST PALM BEACH, FLORIDA 33406

Weird and wonderful...

Nepenthes

Plants of several families have "pitchers" to trap insects. Nepenthes *species have lidded pitchers at leaf ends to entice with an attractive odour and nectar-like fluid on the rim. But down-pointing barbs and waxy walls ensure a one-way trip into a digestive liquid at the bottom.*

Rafflesia

One species of Rafflesia *wins the record for the biggest flower in the world. A metre across, its size is doubly amazing because it is a totally leafless parasite, the huge flowers usually emerging from the liana host's roots but sometimes from high up its stem. The evil-smelling flowers are pollinated by flies, and the seeds dispersed by ants.*

Green Inheritance

THE WWF BOOK OF PLANTS

Anthony Huxley

FOREWORD BY SIR DAVID ATTENBOROUGH

REVISED BY Martin Walters

University of California Press
Berkeley and Los Angeles

University of California Press

Berkeley and Los Angeles, California

Published by arrangement with Gaia Books

This book was written by Anthony Huxley with the help of the World Wide Fund for Nature (now WWF) and the International Union for the Conservation of Nature and Natural Resources (now the World Conservation Union, IUCN).

Editorial	Elizabeth Hoseason
	Juliet Bailey
	Bryony Allen
Design	Bridget Morley
Typesetting	Aardvark Editorial
Direction	Patrick Nugent

This edition first published in 2005.

Gaia Books is an imprint of
Octopus Publishing Group
2–4 Heron Quays, London, E14 4JP, UK

Cataloging-in-Publication Data is on file at the Library of Congress

ISBN 0-520-24359-5 cloth
Manufactured in China

13 12 11 10 09 08 07 06 05
10 9 8 7 6 5 4 3 2 1

Weird and wonderful...

Bucket orchid

The bizarre bucket orchids (Coryanthes) show how to make an insect pollinator look ridiculous! Bees suck attractive fluid on the edges of the huge lips, become inebriated, and fall into the liquid-filled "bucket". They can only escape by squeezing under the flower's sexual organs: it may take a bee thirty minutes to wriggle out – yet it will return and so carry out cross-pollination.

Coco-de-mer

The double coconut or coco-de-mer (Lodoicea maldivica) is the biggest seed in the world, weighing up to 30 kg. Its shape once made it prized as an aphrodisiac, when washed up on far-off beaches. The palm is at risk in its only habitat, one valley on one island in the Seychelles.

For Zoë

The aim of the book is to show, before it is too late, just how rewarding our green inheritance is to mankind – to demonstrate the wonder and worth of plants, and their great potential, to explain why they cannot take any more punishment, and to point out how those that remain can be saved. It underlines the need to save these plants because each plant that becomes extinct represents a loss to ourselves. Plants have fed the world and cured its ills since life began. Meanwhile we have been destroying their principal habitats at a rapid rate.

Now extensively revised and updated, the book was first published in 1984 – in conjunction with the World Wide Fund for Nature (now WWF) and IUCN (now the World Conservation Union) to help launch an international programme for plant conservation.

WWF is the world's largest and most experienced independent conservation organization and is a truly global network, working in more than 100 countries. Its mission is to conserve biodiversity, to ensure that the use of renewable natural resources is sustainable and to promote actions to reduce pollution and wasteful consumption.

The **World Conservation Union** is unique. Its members, from some 140 countries, include 77 states, 114 government agencies, and 800-plus NGOs. Over 10,000 internationally recognized scientists and experts from over 180 countries volunteer their services to its six global commissions. Its 1,000 staff members are working on around 500 projects. This "green web" of partnerships has generated environmental conventions, global standards, scientific knowledge and innovative leadership.

Argyroderma

A number of succulents of the Mesembryanthemum *family resemble the stones of the South African deserts in which they live. This* Argyroderma *is mimicking white or quartzy stones. Other "living stones" are grey, brown, yellowish or green with granite-like speckling.*

Hydnophytum

Hydnophytums *enjoy "mutual benefit" living arrangements with ants. Tree-dwelling species of the South-east Asian jungle, they have fleshy tubers in which ants tunnel dwelling chambers. In return for shelter and nectar, the ants protect the plants against intruders, and transport their seed to new sites.*

Weird and wonderful...

Strelitzia

Birds seeking nectar in the centre of the bird of paradise flower (Strelitzia reginae) stand on a strong trough-shaped platform, unaware that stamens in the trough are pressing pollen onto their feet ready to be transferred to the stigma of the next bloom visited.

Durian

Durian (Durio zibethinus) fruit combines revolting smell with delicious flavour. The smell attracts animals of all sorts who help disperse its seeds – elephants, rhinoceroses, tigers, pigs, monkeys, bears and squirrels – and people who have acquired a taste for it. Durian is mainly pollinated by bats.

Welwitschia

One of the most bizarre plants in the world is Welwitschia. It has two evergrowing leaves. up to 9 metres long, which split to give the impression of many more, and absorb moisture only from sea mists in its native South-west Africa.

Bee orchid

Bee orchids (Ophrys) depend for pollination on deluding male insects into mistaking their flowers for desirable females: they have the right odour, pattern and degree of hairiness. Once alighted on a flower the male insects attempt to copulate.

Contents

Foreword

Jade vine

The spectacular jade vine (Strongylodon macrobotrys), almost unique in flower colour, is severely endangered in its native habitat, the Philipines. Fortunately, it readily takes to cultivation in tropical gardens or under glass, as in the Palm House at Kew, England.

Butterwort

Insect-eating butterwort (Pinguicula) invented fly-paper before we did. Small insects alighting on this bog plant's sticky leaves are stuck fast. Glands on the surface then exude more sticky juice, and an acid fluid which dissolves the prey. Meanwhile, the edges of the leaf curl slowly inwards, hindering escape.

The need for this book has never been greater. Never has it been more urgent or more important that the message it carries should be widely heard and understood. For what it makes marvellously and soberingly clear is the extent of our dependence upon plants.

Plants provide us with power. In many places, people burn them to cook their food and to keep themselves warm. Even in industrialized parts of the world, fossilized plants, in the form of coal and oil, give us most of the energy with which we run our machinery and illuminate our cities. Nearly half of all our drugs are based on substances that are derived from plants. Plants maintain the level of oxygen in the very air we breathe. And who can quantify the joy that their beauty brings to our spirits?

Indeed, animals could not exist without plants, for all – including humans – rely upon them for food. Even the lion that lives on antelope is eating plants, as it were, second-hand. As the Bible says, all flesh is grass. Putting that truth into different words,

plants alone have the ability to combine atoms of carbon, oxygen, hydrogen and nitrogen and produce the complex molecules that are the building blocks of living tissue and the essential food of animals. So plants not merely sustain all forms of life, but continuously re-enact the ultimate and fundamental miracle of turning the non-living into the living.

Obvious though our dependence upon plants may be, we have, it seems, taken it for granted that they will continue to exist in all their variety no matter how we treat them and the land on which they grow. The initial concern of many conservation organizations was to protect rare animals. Maybe it was the very mobility of birds and mammals that first attracted our attention to them and diverted our care from the green organisms that remained stationary, rooted to the ground. But slowly, people began to realize that the survival of the animals about which they were concerned depended on the continued existence of plants – and, nearly always, upon particular kinds. If you want the yearly delight of butterflies in a garden, you must grow special things – buddleias and nettles, honeysuckle and bramble. And if you want to keep giant pandas, you have to grow bamboo!

Initially, the main problem seemed to be with preserving the butterfly and not the nettle, the panda, not the bamboo. Many plants, after all, can be kept as seeds in a packet, transported from one site to another by post if needs be, and brought to life by a little judicious watering. Surely they could not be endangered too. Now, to our cost, we know better.

No one is certain exactly how many different species of plant exist. A reasonable guess is that there are at least a quarter of a million flowering plants alone. Two-thirds of these grow in the tropics. Yet it is precisely there that their destruction is proceeding the fastest. The overall global rate of forest loss has slowed in recent years, because losses in some areas are offset by planting and natural growth elsewhere. Nevertheless, net losses still run at some 9,000 sq. km a year – an area the size of Hungary or Portugal – and deforestation is severe, especially in tropical forests. Of the 95,000-plus species of plants growing in South America, only about one percent have been examined to see what value they might have as food, or for medical or any other purpose. So, year after year, we are destroying species without even knowing what we are losing. All over the world, we are draining marshlands, felling forests, ploughing up grasslands, and adding damaging chemicals to the soil and the air. Worldwide at least 31,000 species are now officially recognized as threatened, but the true figure is certainly much higher; indeed, some estimates put this at over 75,500, and possibly as high as 94,400.

Having, in 1984, launched a worldwide campaign to save the plants that save us, WWF and the World Conservation Union (IUCN) continue to make a sustained effort to halt this devastation.

The illustrations in the book have been selected from, among other sources, the rich library of drawings assembled by the premier botanical gardens in the world at Kew, in England; many of its facts have been supplied by research scientists belonging to the IUCN; and the whole was put together by Anthony Huxley, a botanist who was as knowledgeable about the nature, distribution and value of plants as he was skilled in actually growing them, whose breadth of vision was matched by his invaluable ability to express the findings of his science in the simplest of words. You will not find here strings of those baffling adjectives – glabrous and glaucous, oblanceolate and indehiscent – to which so many botanists are attached and which so many non-botanists find such a barrier to comprehension. In chapter after revelatory chapter, Anthony Huxley unveils the beauty and wonder of the plant kingdom and shows just how intimately our lives are bound up with it. Perhaps his wisdom and his vision will at last persuade us to set about the task of protecting what he describes with such clarity. It will not be before time.

Cycad

Half tree, half fern in appearance, cycads like Cycas revoluta *are in danger from collectors, as ornamental plants. Known as fossils from 200 million years ago, their life span measured in thousands of years, these ancient plants grow only from seed.*

Ceropegia

Ceropegia haygarthii *waves a hairy club-shaped organ to attract insects. Odour then leads them to the slippery chute of the flower proper, which restrains them until their pollinating work is done, when the bloom tips to release them.*

CHAPTER ONE

The Green Inheritance

Plants are very familiar. They tend to be taken for granted and regarded with placid affection rather than with the fascination and interest they deserve. For most people they are part of the view, whether open country, farmland, or townscapes with street trees. Yet they are unique among the organisms with which we share this planet, for they alone contain the pigment chlorophyll which allows them to derive their energy from light. Outwardly, "They toil not, neither do they spin", but inwardly they are highly active and amazingly complex chemical-producing factories.

There is more than one view of plants. To the more comfortably off they include garden or park plants for leisure and pleasure, and the indoor pot plants which provide for town dwellers that contact with the green wild world that seems to be an instinctive need.

By contrast, many people in developing countries depend heavily on plants, but look at them hardly at all from the standpoint of beauty or solace. Crops are there to be cultivated, often with extreme sweat and toil, and wild plants to be exploited directly. The overwhelming need for fuelwood and grazing for animals is often totally unselective.

We use plants in every field of life and have long ceased to rely on those native to our own area. The average Northerner has foods, products, and material contributed by plants from all over the world, many grown far from their original homelands – exotic hardwoods from the tropics for flooring and garden furniture, fruit and vegetables from around the world, insecticides from chrysanthemums, medicine from the Himalayan opium poppy or the Andean coca, spices from the east, dyes from Asia, polishes from the jungle, toiletries from jojoba, grown in the desert; pot plants from the tropics, oils from the Mediterranean olive; soya from Brazil; cotton, sisal, rubber – the list is almost endless.

Not very long ago, the plant world seemed inexhaustible, always reasonably renewable. Today it is all too clear it is not. Nevertheless, we continue to wrench plants from the ground or destroy their environment and, whether for reasons of pressing immediate need or for greed, eradicate natural plant life as we do so. It is a sad fact that few people, whether the local tiller of soil or the modern entrepreneur, have any real regard for the world of plants, either for the plants' sake or for their own. The wild plant is in deep trouble; there is no mistaking that.

Numbers and niches

Plants often surprise people by their diversity. Botanists have been describing and cataloguing plants for over 200 years, but we still do not know how many species there are. Some recent estimates have put this figure as high as 422,000 species of flowering plant. There are many more plants, in fact, than higher animals: the vertebrates – animals with backbones – consist of 46,000 species of which about half are fish, over 4,600 mammals and 10,000 birds. But in contrast there are far fewer plants than invertebrates, of which a staggering 1.4 million have been described and many more await discovery.

The crater lake of Maninjau, in Sumatra, is a famous tourist site. The deep lake is flanked by wooded groves, partially cleared and terraced for growing rice, as here on the eastern shore.

The number of plant species in the world can only be an estimate rather than a known total. As new areas are explored, botanists continually find plants never seen before; one expert believes that there may be about 10,000 undiscovered plants in Central and South America alone. Indeed, for some parts of the world, knowledge of plants is still at a rudimentary level. Botanists all over the world are working feverishly to catalogue and describe plants; as vegetation disappears it has become a race against time. In fact, whereas most of the plants grow in the tropics, most of the botanists live in the temperate realm where plant life is much less diverse and thus better known. However hard botanists work, many plants will inevitably become extinct before they have been seen and described, let alone saved for the future and screened for useful products.

There are two main groups of higher plants: the angiosperms ("enclosed seeded") or flowering plants; and the much less numerous gymnosperms ("naked seeded") which include the conifers and the more primitive, palm-like cycads. The latest estimates put the number of species of flowering plants at about 422,000, and the gymnosperms at about 840.

In addition, there are about 12,000 species of ferns, horsetails and clubmosses, and some 16,000 mosses and liverworts. The fungi, now classified in their own separate kingdom, comprise about 77,000 species, with many more to be discovered and named. The algae are a very diverse and varied group, and contain single-celled and multi-celled species from microscopic forms to seaweeds, some of which are huge. About 40,000 species are known, but there may be as many as 300,000 species. The marine algae form an integral part of the sea's life system, providing home and nutrition for huge numbers of organisms, and releasing no less than 70 percent of the world's atmospheric oxygen.

Some seaweeds are extraordinary, often having fronds 15 metres long, while the giant kelp, *Macrocystis pyrifera*, is reputed to reach 200 metres long, including its deep-fastened stem.

It is fascinating how all the stages in plant evolution – the different groups which reached standstill at various points on the evolutionary road before the flowering plants appeared – now exist together on the land surface, each occupying niches according to their capabilities. These often overlap – both ferns and orchids, for instance, co-exist as epiphytes on the trees of the tropical rainforest.

Which plants grow where is controlled by many factors including climate, temperature, light, water availability, soil composition and acidity.

The botanist classifies the world's vegetation into a range of zones which reflect the local climate. In each a unique set of relationships is built up, the plants always forming the basis of each life system or ecosystem, supporting all its diverse inhabitants. In few cases could animal life of any kind exist without plants, even temporarily.

In the bleak tundra, below freezing for much of the year, lichens and algae, covering bare rock surfaces, are the only plants close to the ice caps; they also occur on high alpine rocks beyond the level of flowering plants. But in both arctic tundra and alpine zones below the bleakest levels, grasses, sedges, herbaceous plants and dwarf shrubs thrive; the flowering plants demonstrating their adaptation to the environment by creeping flat on the ground or in cushions, with roots often extending widely in shifting scree or permeating rock crevices.

Throughout the different zones one can see plants specially adapted to each set of circumstances. In hot, moist climates tree leaves are often large. In cold climates, leaves are usually small, needle-like and evergreen, losing their moisture very slowly by having thick cuticles and pores (stomata) in deep pits. These occur, for example, in the conifers of very cold forest zones with short growing seasons and in the characteristic heath forest of some tropical areas where soils are very poor and drain very freely, so that water conservation is equally essential. In Mediterranean climates, where the summers are hot and dry, leaves tend to be evergreen, small and leathery.

Some of the most obviously remarkable adaptations are found in arid semi-deserts. The cacti, almost all American, and the succulent spurges from Africa, have mostly abandoned leaves, transferring their duties of photosynthesis to the stems which have become enlarged into the typical sphere and cylinder shapes, those in which most water-retaining volume can be fitted into the least surface area capable of losing moisture by evaporation. Cacti and spurges form an example of parallel evolution – plants ancestrally quite different becoming

*The Olympic peninsula at the north-west extremity of the United States has the most northerly example of temperate lowland rainforest, **below**, with average annual rainfall of 350 cm. Tall conifers and lesser deciduous trees are draped with mosses, clubmosses and ferns which thrive in the very humid conditions.*

*Many landscapes which we think of as natural have in fact been modified by past human activity. This New Hampshire woodland of oaks and maples, **above**, has a different – poorer – mixture of species from its pristine composition because of selective cutting over the centuries.*

*The characteristic high-altitude vegetation of the northern Andes (over 3,000 metres), called paramo, **above**, is composed of bunchgrass, sometimes with small bamboos, many small-leaved shrubs, and rosette plants like the espeletias which dominate this Colombian view. This natural plant complex can be compared with that of European mountains, **left**, which are often modified by the grazing of domestic animals that are moved to the high grassy alps in summer, while in the lower reaches the lush meadows are cut for hay. However, the wild plants continue to thrive under these circumstances, in total contrast to terraced rice paddies as in Nepal, **far left**, where the original vegetation has been entirely removed and the landscape transformed.*

superficially similar because they live in similar environments.

Most other succulents have not lost their leaves but have developed them into comparable swollen shapes. In one family, the mesembryanthemums of southern Africa (*Aizoaceae*), one can trace a whole range of genera in which the fleshy leaves grow closer and closer together, finally reducing in number to three or four tight-packed pairs, a single pair, and a coalescent pair making a "plant body" as in the stone mimics (*Lithops, Conophytum* and so on). In this way the character of having separate leaves is transformed to resemble rounded cacti.

Each vegetation zone has its special needs and its unique set of relationships between different kinds of plants and between plants and animals. The most complex is undoubtedly that of the tropical rainforest which originally girdled the earth around the equator. Here there are often no marked seasons, and days and nights are of roughly the same duration. Rainfall is high, often coming in prolonged downpours, and fairly evenly spread through the year. Rainforests usually exist on a very shallow humus layer, almost all the vegetable matter – the biomass – being in trees and other plants. Among the trees, which occasionally reach a height of 60 metres, the variety of niches ranges from low shade-tolerating plants on the ground to the lianas which clamber towards the light, devoting minimum energy to supporting mechanisms, and the epiphytes which cluster on upper branches with no connection with the soil, clinging by aerial roots which absorb moisture from the air. Many larger-leaved plants are covered with a scurfy film of minute hanger-on plants called microphylls.

Distribution and diversity

Plant species are unevenly spread over the land. In general terms there are far more in the tropics than in temperate regions. More than 90,000 occur in Central and South America with 56,000 in Brazil alone; tropical Africa accounts for at least 30,000, and the island of Madagascar another 9,500. As one might expect, going north towards the Arctic reduces the number of species; the British Isles have only around 1,800, though Mediterranean Greece, five-eighths of its size, has over 5,000. Greece has a high proportion (about 14 percent) of endemics – plants found nowhere else – but perhaps the highest ratio is found in the Hawaiian Islands where more than 90 percent of plants are endemic. A great many plants are very local indeed, especially on islands or isolated mountains where independent evolution has continued, and these are often the most endangered.

Many such plants are, not surprisingly, highly adapted to a particular environment, but this does not always mean that, if transferred elsewhere, they will not grow happily. Many maligned plants have spread widely after transfer, usually as a result of being introduced to gardens as ornamental plants, like Bermuda buttercup (*Oxalis pes-caprae*), prickly pear (*Opuntia*) and water hyacinth (*Eichhornia crassipes*), none of them particularly strong colonizers in their original habitats. Some introduced colonizers swamp and destroy native vegetation – Hawaii is in process of having many of its endemics choked by the invasive lantana (*Lantana camara*): nearly 300 plant species are threatened, many reduced to a handful of individuals. Others spread rapidly as weeds of cultivation. Many such weeds can be said to be gifted with an innate ability to take advantage of the disturbed areas that follow cultivation, road building and the like. Such adaptability characterizes other plants which are naturally widespread cosmopolitans, like the fern bracken (*Pteridium aquilinum*) – one of the most powerful invaders of all kinds of terrain. A bracken plant seems immortal and immeasurable, spreading as it does from underground growth buds at a rate of one metre a year.

Since the world is full of vastly differing habitats for plants and climates in which they grow, it is not surprising that plants exhibit an equally vast range of form. Within the gymnosperms and flowering plants alone they range from the minute duckweed under a millimetre across, to huge forest trees. The tallest living trees are the coast redwoods (*Sequoia sempervirens*) of northern California and southern Oregon, USA. They often exceed 100 metres, and the current record-holder is the stratosphere giant, now 112.6 metres tall. Australian mountain ash trees (*Eucalyptus regnans*) regularly grow to over 95 metres, and the tallest reliably recorded was 112.8 metres when felled in 1884.

Flower size varies from the fraction of a millimetre of the duckweed to the near-metre of the parasite *Rafflesia*; seeds range from those of epiphytic orchids which weigh in at 1,000 million per gram to the double

Barley is one of the more remarkable of our cereal crops because, though originally a plant of arid lowlands in the Near and Middle East, it will thrive in mountain areas, as here in Nepal, **above**, into the Arctic, and even in saline desert conditions, often in places where other crops refuse to grow because of drought.

When unaffected by severe drought and both animal and human overpopulation, the African savannah, as seen in Tsavo Park, Kenya, **top**, provides a perfect environment for a wide variety of creatures large and small. Dominated by acacia trees, the surrounding grassland becomes lush in the rainy seasons and dries up in between.

The Joshua tree, **above**, a branching species of yucca, is characteristic of many North American deserts. Unlike the cacti with which it often grows – which have lost their leaves and developed fleshy water-retaining stems – the Joshua tree protects itself against drying out by well-insulated, spiky leaves.

coconut (*Lodoicea maldivica*), which can weigh 18 kilograms.

Even more remarkable are plant construction, metabolism, and specialized devices. Consider for instance the engineering which enables a tree to hold itself up and withstand all but the fiercest winds; the internal plumbing system that ensures that every part of the plant from the top down to the roots receives products of the photosynthetic energy generated in the leaves and, in reverse, that the leaves receive supplies of water, essential for photosynthesis, from the roots; the force needed to pump water up a 100-metre tree is enormous.

Every plant is a chemical factory for complex substances which exceeds any human capability; even a bacterium can outperform us in this respect. In their seeds, plants invented aerofoils aeons before the Wright Brothers; in their bulbous roots, they were well ahead in devising winter larders. In their poisons, antibiotic agents, prickles and foul tastes, they developed defences against attack long before human stockades and pesticides. Their anti-exposure devices are manifold. The typical insulation is of hairs, often dense and furry, of waxy cuticle, or layers of lifeless, air-filled cells under the surface. The extraordinary giant lobelias of central African mountains have long woolly leaf bracts which fold over the developing shoot or flowerhead at night. In addition, the shoot tip is itself surrounded by a liquid; at night the top 2–3 cm of this freezes, but around the shoot tip the liquid remains just over freezing point. The nearby tree groundsels protect their trunks by retaining their old leaves, as a dense layer of lagging.

Bark is the standard insulator for trees, sometimes thick and corky, as in the cork oak (*Quercus suber*), to combat heat and dryness, or in the Hawaiian ohia lehua (*Metrosideros polymorpha*) which resists red-hot cinders from volcanic eruption. A dramatic example is giant sequoia (*Sequoiadendron giganteum*) from California: its branches start up to 60 metres from the ground, so when forest fires rage below, its 60 cm thick bark keeps it unaffected.

It is because plants cannot run away from their predators that they have had to develop convincing deterrents: all manner of spines and thorns; the stinging hairs of nettles which resemble a hypodermic syringe in action; saw teeth on leaves; bitter or blistering sap, and leaf cuticles reinforced with abrasive silica. A few plants protect themselves by mimicry, notably the many kinds of "living stones" (*Lithops*) in South Africa, resembling the pebbles in which they grow, brown, grey, or white and quartzy. Other succulents, like some *Anacampseros*, look like nothing more than bird droppings.

Some plants have taken to eating meat: butterworts (*Pinguicula*) and sundews (*Drosera*) have sticky surfaces or tentacles which trap and enfold insects. Venus's flytrap (*Dionaea*) has a pair of spine-edged pads with sensitive hairs in the centre – an insect tripping them is instantaneously shut in and then digested. *Nepenthes* and *Sarracenia* have pitchers with an attractive odour, barbed hooks and slippery edges at the top, and a reservoir of digestive fluid at the base. All these plants live in nitrogen-deficient habitats like bogs and need flesh to supplement their diet.

The movement of the Venus's Flytrap is one of the fastest in the vegetable world, but many plants can move some of their organs in response to stimulus. Most familiar is the sensitive plant (*Mimosa pudica*), whose leaves and stems collapse at a touch. Plants with tendrils move these around supports very rapidly when contact is made. To aid pollination, some plants have moving stamens or other devices in the flower.

Pollination provides the most numerous and amazing, highly tuned devices for ensuring that an insect, bat, bird or, sometimes, small mammal transfers pollen from one flower to another. Odours are the most obvious bait – not just the sweet ones which typically attract bees and moths but, to us, odious ones for insects like flies. Stapelias and their kin smell of decay and their lurid red, brown and yellow tones resemble rotting meat; some even have hairs to mimic mould. Flowers that attract bats for pollination have strong odours, typically musty, fishy or foxy.

The orchids are the most highly adapted plants of all in this respect, one species usually pollinated by a single species of insect. The huge lip of the bucket orchid (*Coryanthes*) is, literally, a squarish bucket with a trough-like spout filled with liquid from special glands. Its edges are edible and highly attractive to bees, which become inebriated; they fall into the bucket, from which the only escape is under a rod-shaped organ bearing pollen on its stigmatic surface. The Mediterranean bee orchids, *Ophrys*, are pollinated by male wasps because they smell, look and feel like female wasps – the males "pseudocopulate" with

An extreme of the evolutionary process is shown by the many plants which parasitize others. The dodders, which infest large trees in the tropics, germinate normally in soil, but once they have found a host, plunge sinkers or "haustoria" into its tissues and lose their own roots. Their nourishment is derived from the host, their own leaves being reduced to scales.

The African succulent stapelias are called carrion flowers because they simulate rotting meat in order to attract the blowflies which pollinate them. They have the lurid colours and odour of decay, and many carry fine hairs which move in air currents and mimic moulds. Flies are often so thoroughly deceived that they even lay their eggs on the blooms.

The leaves and stems of the sensitive plant (Mimosa pudica) collapse at a touch at a rate of about 3 cm per second from the point of contact – perhaps to discourage browsing animals, which see their lunch disappearing as they brush it. At night the plant folds its leaves into a "sleep" position – a habit shared with a number of other plants, the purpose of which is unclear.

The leaves of the giant water lily, Victoria amazonica, **left**, are a wonderful example of plant engineering. Underneath they have an array of girder-like radial and transverse ribs which keep the two-metre-wide surface rigid and, since they contain air spaces, afloat. A big leaf can support a child's weight. Large prickles on the underside probably deter predatory watersnails.

the orchid, deceiving themselves but achieving success for the flower.

Plants which serve us

Plants are remarkable, then, in their own right – as objects for study, to amaze us the more we learn about their lifestyle. Many are prized for their beauty and cultivated accordingly; others have innumerable utilitarian applications which technology shows no sign of superseding.

Every year, our farmers have to find food for more people; our forests have to provide more timber and fuel; our chemists have to supply more medicines; our industrialists need more raw materials to meet human needs and demands. Unless the population stabilizes, which on present projections may not happen before the twenty-second century, and some kind of equilibrium is reached, this battle to squeeze ever more out of the living world will intensify year by year. And right in the firing line are the plants – the green inheritance on which we, and all animal life, depend.

Some plants are pre-eminently valuable. The cotton plant (overleaf), *Gossypium*, is a striking example of the "multipurpose" plant, and other members of its family, *Malvaceae*, demonstrate how many surprising uses they have. This, the mallow family, includes familiar garden plants like hollyhocks (*Althaea*), hibiscus, and various mallows. A great many of these have strong stems which produce fibres of various qualities, some fine, some as strong as jute. They contain mucilaginous substances which make them valuable in soothing medicines

(hollyhock flowers, for instance) and also in sweetmeats of various kinds; marshmallow is an ancient confection which used to be made from the root of *Althaea officinalis*.

We easily forget how often we use plants. Cotton is just one example. In Norman Myers' words, we benefit from plant materials every time we apply a shampoo or sunscreen lotion, paint a wall or varnish a table; every time we use a golfball or jet engine or oil-drilling equipment; every time the dentist makes a mould of our teeth or the doctor gives us a vaccination. The plant kingdom is truly a part of our industrial and social jungle.

The web of life

The richest web of life, the greatest concentration of species, is found in the world's rainforests like those of Amazonia. Plants are the rainforest's fabric: a great assemblage of trees, ground-living plants, epiphytes perched higher up, and lianas bridging the gap between floor and crown.

In the dim world under the layers of vegetation, insects are to be found everywhere, from the microscopic to the great cockroaches and brightly coloured beetles; often of weird shape or mimicking leaves, twigs, thorns, seeds or bird droppings. Bees and fierce wasps make large colonial nests. Others like the mantis prey on their relatives, while spiders lurk at every point and level, trapping, snaring or pouncing and culminating in the enormous hairy bird-eaters. Apart from midges and mites, which can madden the human visitor, butterflies are prominent – great blue morphos flitting

slowly, others shaped more like dragonflies, some with almost transparent wings. Their caterpillars, and those of moths, feed on the leaves; their adults help to pollinate the trees' flowers.

Ants are everywhere, from large, aggressive army ants to smaller leafcutters carrying pieces of leaf from selected trees for hundreds of yards in endless streams, to underground caverns where they grow fungi on leaf compost to eat. Termites (similar in appearance to ants but biologically closer to cockroaches) are concealed in large nests of chewed wood and saliva up the tree trunks, travelling to soil level via tunnels to feed on debris. Together with the innumerable fungi, they are the most important agents of decay in the forest and are responsible for the release of plant nutrients back into the system.

In their turn insects are meat for small rodents, lizards, frogs and birds. Birds are surprisingly difficult to see but they are there in plenty; puffbirds and jacamars catching butterflies on the wing, and antbirds on tree trunks or on the ground, while hummingbirds suck nectar high up in the flowering canopy. Fruits and nuts are consumed by huge-beaked toucans, gaudy macaws and other parrots, all adapted with short wings for inter-tree flying and equipped with pincer-like feet for clambering. In the branches there may be primitive, ungainly hoatzins, a bit like pheasants, which eat aroid leaves; heavy fruit-eating curassows which clamber up trees before gliding down again; and, on the ground, flocks of aptly named trumpeters. With extra luck one may see a number of brilliant gold

To most of us cotton is just a fibre, useful for all kinds of thread, clothing and fabric. About 95 percent of world cotton production is taken from the seed fibres of Gossypium hirsutum – *a medium-long-staple species originating in Central America. The extra-long-staple G.* barbadense, *from South America, is shown being harvested here. The short-staple G.* herbaceum *and G.* arboreum, *less widely grown, come from Africa and Arabia, and India and Pakistan respectively. The fibre has many uses: for cordage, carpets and rubberized fabrics, and as cotton wool, used both in surgical dressings and cosmetic cleansing. The wool is the sterilized fibre cleared of its natural oiliness.*

And fibre is not the limit of the plant's usefulness. The seeds are rich in oil for cooking, salad oils and lard substitutes, for margarine, soap and soap powders. The oil cake left after the oil has been pressed is valuable for cattle fodder, and fertilizer, but also as a basis for explosives and alcohol, and in lining oil wells. The stems, if not burned for fuel, are fibrous enough for paper-making. The root bark has medicinal properties including that of stemming bleeding. The flowers are a good source of honey and, in India, a yellow or brown dye is obtained from the petals.

Even today, the cotton plant may not have exhausted its repertoire: its seeds may be able to supply us with the elusive "contraceptive for men" (p. 122).

male cock-of-the-rock leaping and bobbing to attract mates.

Bats are mainly fruit and nectar eaters. The teeth and muscular tongue of the bat *Corollia* enable it to crush every bit of goodness from a fruit and then to spit out skin and seeds, thus helping dispersal. Pollination is aided by such species as the long-tailed bat *(Choeroniscus)* which sucks nectar from flowers: bat-encouraging blooms are large and sturdy, as indeed they must be to stand up to the weight of the bats hooking on to them to feed.

Birds of prey, notably the spectacled owl and the huge harpy eagle, mainly attack the tree-inhabiting mammals. Most spectacular of these are the monkeys – small chattering capuchins, woolly, spider and well-named howler monkeys, all with prehensile tails (unlike monkeys in most other parts of the world) which enable them to spend almost all their time in the branches. Here too are furry little kinkajous, feeding on fruits or honey, and tree porcupines eating fruits and leaves: both have the invaluable prehensile "fifth limb".

High in the trees live the sloths, hanging upside down and motionless much of the time, camouflaged by green algae growing on their damp hair; they come to the ground only to defecate. The related dwarf anteater is almost as sluggish. Like the larger tamandua it has a prehensile tail and feeds by scratching out ant and termite colonies.

Many warm-blooded creatures forage on the ground – small deer, tapir, peccary, giant anteater, and the herbivorous rodents, agouti and paca. These search variously for roots and tubers or for the multifarious life in the leaf litter – insects, molluscs, worms and crustaceans. Coatis, their groups up to forty-strong, are omnivorous, eating animals and fruits, and scrambling in the lower branches as well.

These foragers, as well as birds, lizards – anything for a mouthful – are sought by carnivorous cats – ocelot, margay and jaguar, this last dropping silently out of the trees onto its prey. Jaguar will also fish, using its tail as a lure. Tayra and grison also attack; they are sometimes tamed by Indians to flush out agoutis. Packs of the stumpy bush dog hunt the large rodents too.

The giant anaconda is yet another hazard to mammals, while the emerald tree boa eats smaller prey such as young birds. Both kill by strangulation; venomous snakes are few. Smaller kinds creep up the trees, pouncing on insects, lizards and frogs.

Frogs are mostly small and often brightly coloured, like the viridian arrow-poison frog which, after its eggs hatch, transports its tadpoles on its back to the safety of the water reservoir of an epiphytic bromeliad. In these "urn plants" or "tank epiphytes" there is indeed an amazing miniature ecosystem. Tiny crabs and mosquito larvae feed on protozoa in the water and dragonfly larvae feed on these. Carnivorous bladderwort plants often establish themselves in the water reservoir. Beside the frogs – some of which spend their entire lives within one plant – snails, worms and salamanders appreciate the moist shelter, and these are prey for birds and mouse opossums as well as large insects attracted by the water.

It is a sobering thought that when the jungle is destroyed hundreds of species of birds, bats and large mammals, and tens of thousands of insects and other small creatures, disappear forever. Without the ambience of the forest they cannot survive.

The jungle took millennia to develop; a tree can be destroyed in minutes. Important and fascinating as it is, however, the jungle is but a part of our green inheritance. This – the genius of evolution and plant survival strategies, and the genius of thousands of human generations – is the capital on which we live today. The bowl of rice on your table is a gift of the ancestral wild rice, of the centuries of rice growers, of the skills of modern botanists, of the disease resistance of a wild relative recently "borrowed", of the work of soil micro-organisms. What should we do for rice today if the first wild rice ancestors had been ploughed under by a bulldozer to make way for a Neanderthal holiday resort?

We have loved plants, and sometimes worshipped them, all through history. Far more than animals, plants have been symbols of the earth's life force and have provided solace to the human race. Yet we are squandering this inheritance in ignorance, in thoughtless impatience and greed, failing to appreciate either the beauty or the value of what we destroy. The aim of this book is to show, before it is entirely too late, just how rewarding our green inheritance is to mankind, what potential it has, how it cannot take much more punishment, and to seek ways to save what remains.

CHAPTER TWO

Guardians of the Environment

Plants are continuously active, by day conjuring the food they need out of sunlight, water and gases around them, like magicians pulling rabbits out of a hat; by night and day, using the stored energy for growth and life. This miracle of living alchemy is called photosynthesis. Without it, virtually all life on this planet would cease.

In essence, a plant taps the energy of sunlight to make carbohydrate out of carbon dioxide gas and water. The plant then uses this carbohydrate for energy, and as a building material to synthesize proteins, fats, and all the complex chemicals of life, drawing also on minerals in the soil.

Plants are self-sufficient, able to colonize new terrains and survive by photosynthesis, where animals could not. They were the first organisms to emerge on dry land, in the course of evolution. In the Arctic, the algae are the first colonizers of rocks and debris, their slime creating a toehold for other microscopic organisms. Multicoloured single-celled algae grow on snow; at the other extreme, cyanobacteria (sometimes wrongly called "blue-green algae") can live in temperatures of up to 35°C. Plants, too, are the first recolonizers of damaged, hostile environments, from devastated wasteland to natural salt land and lava flow. Of all the benefits plants can bring, perhaps the greatest is the potential to repair the harm we have already done.

Plants produce food: animals consume it. Without plants, animal life could not exist. The range of plants, from single-celled algae to great forest trees, is food for vast numbers of creatures in the sea and on land; these creatures are themselves food for carnivores.

Hot sulphurous springs in Yellowstone National Park are colonized by brilliantly coloured bacteria.

This is the food web we all understand, with humanity at its centre, omnivorous but relying to a large degree on food derived directly from plants. At each point up the chain, there is a great drop in the number of organisms: one whale depends on millions of krill; one person eats the bounty of perhaps millions of plants and thousands of animals in a lifetime.

Plants provide the fabric of our land-scape, the home for all animal life. In the era before human dominance, natural vegeta-tion flourished as a green tapestry over the earth, endlessly varying from tall forest to waving grassland, from tropical savannah to arctic tundra, reflecting changes in climate, soil and terrain. It formed a natural security blanket on the surface of the earth, helping to maintain the balance of gases in the air and nutrients in the soil, stabilizing climate, river flows and soil cover.

Guardians of air, climate, and soil

We have always taken for granted these services of plants in maintaining the "life-support systems" of the planet. Plants are life-givers: it was the emergence of green plants, long ago, that gave us the air we breathe – the primeval atmosphere would poison present day organisms. Our oxygen-bearing atmosphere was a by-product of early photosynthesis, mainly by marine algae; these ocean plants still provide the bulk of our oxygen. All plants are involved in maintaining constant oxygen and carbon dioxide levels, their ceaseless exchange of these gases playing

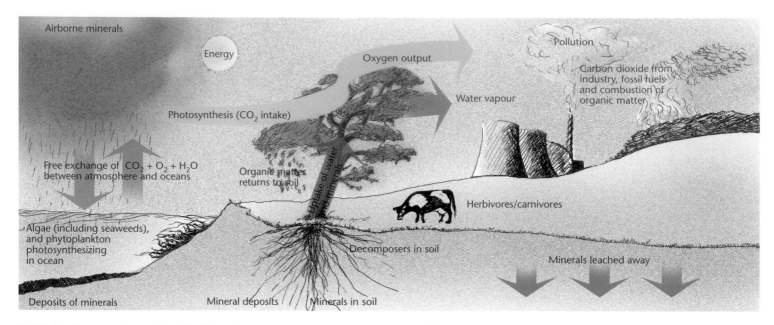

Airborne minerals

Energy

Oxygen output

Pollution

Carbon dioxide from industry, fossil fuels and combustion of organic matter

Water vapour

Photosynthesis (CO_2 intake)

Free exchange of $CO_2 + O_2 + H_2O$ between atmosphere and oceans

Organic matter returns to soil

Uptake of water and minerals

Herbivores/carnivores

Algae (including seaweeds), and phytoplankton photosynthesizing in ocean

Decomposers in soil

Minerals leached away

Deposits of minerals

Mineral deposits

Minerals in soil

While busily converting carbon dioxide and water to carbohydrate, plants give out oxygen, a by-product to the plant but essential to all human and animal life. (Interestingly, about half of the oxygen in the atmosphere is produced by sea algae, mostly microscopic plant-plankton.) They also catch carbon, returning some to the air, but passing some into food chains, and also into the soil.

Another vital element, nitrogen, makes up almost four-fifths of the air. As a gas, however, nitrogen is unavailable to plants. It is brought into their life cycle mostly by microbial organisms; some of these fix nitrogen and synthesize nitrogenous compounds in symbiotic association with certain plants – mostly legumes. When such plants and bacteria die, other bacteria convert their protein to compounds of ammonium, which can be absorbed by plant roots from the soil solution to build up their proteins.

In addition to nitrogen, plants take phosphorus and potassium from the soil – the N, P, K of fertilizer packs – as well as calcium, magnesium, and trace elements. These then find their way into the food chains and, via decomposers, back to the soil.

its part in the global cycling of gases between air, ocean, rocks and soil.

Throughout evolutionary history, temperatures on our earth have remained generally favourable to life (if less comfortable during ice ages). The level of carbon dioxide in the air has provided one form of temperature control; the gas acts as an insulating blanket, trapping much of the heat energy radiated from the earth, which would otherwise be lost to space. Water vapour in the air is important too. Plants assist in the cycling of water through soil and air, transpiration from their leaves providing humidity which encourages rain – over the great moist tropical forests, clouds can be seen gathering, to stop short where the lush greenery ends. Plant and cloud cover affects the earth's "albedo", a measure of its shininess or tendency to reflect sunlight (and heat) back into space – another temperature control.

Today, the world's green mantle is disappearing fast; forests are felled, erosion spreads. One has only to enter a barren area to realize what the immediate consequences may be. In plantless lands, the sun beats down, reflecting back from the bare ground; at night, temperatures plummet,

since no heat is retained. The air is extraordinarily dry, and so is the soil; it rarely rains, and when it does the unprotected soil is rapidly washed away. Such conditions are inimical to all but the most specialized forms of life. Although many deserts are natural and some are very beautiful, loss of vegetation, through over-grazing and other human pressures, is more and more often a prime cause of desertification. As plant cover goes, climates change and soil is eroded.

If the forests are lost, a link in the carbon cycle is weakened because, in their photosynthesis, plants catch carbon. Carbon dioxide levels in the air are rising; since 1970 they have increased from about 310 to 380 parts per million, and may top 400 ppm in the next decade. (Combustion of fossil fuels and changes in land use through large-scale loss of forests are the main sources.) It is accepted that concentrations of carbon dioxide and other greenhouse gases such as methane, CFCs and nitrous oxides are increasing at a rapid rate.

Most scientists now agree that this will lead to human-induced climate change, with an increase in global mean temperature of about 2°C by 2015 and an associated sea-level rise. The world's climate zones will also shift, perhaps at a rate faster than the ability of many species to adapt. As a result of this shift, vegetation and hence landscapes will change, the current food-producing areas of the world may be less productive, and areas already suffering drought may deteriorate further. Low-lying coasts will be inundated by sea-level rise, threatening cities and agricultural lands.

The effects of a rise in the earth's temperature cannot be predicted with any certainty as the interactions of the world's oceans, clouds and major ecosystems and climate have not been fully assessed. For example, the release of large quantities of methane from tundra areas as they warm could create feedback, further adding to global warming.

We know too little about the incredibly complex, interrelated mechanisms of atmosphere, climate, and plants to resolve such conflicting theories. We do know, however, that human modifications of plant cover already have significant effects. Cutting down forest in Brazil not only reduces rainfall locally but also affects the unnatural spread of deserts in nearby Peru. How far the effect reaches is not known, but some concern is expressed for the US grain belt. We should be very cautious in what we do.

Plants stabilize the atmosphere and climate: they also create and protect the soil, perhaps our most precious resource. Soil is a mixture of mineral particles and humus. Mineral particles are largely made available by weathering of rock by heat, cold and water, especially acidic or fast-running water; plants help this process only in a small way. Humus on the other hand results from the

decomposition of organic matter (both vegetable and animal). It is jelly-like, coating the soil particles, swelling when wet and shrinking when dry; it stabilizes soil structure, holds water, but allows air to percolate. It contains all the soil's reserves of nitrogen and retains many other essential nutrients.

Roots of plants both secure them in position and absorb nutrients; questing through the earth, they create channels valuable for drainage and aeration, and may tap minerals and water from layers of rock below. The amount of minerals required by different plants varies enormously. In the rainforests, epiphytic orchids, bromeliads and ferns live high on the trees, with very little mineral supply at all; cacti and succulents, by contrast, inhabit semi-deserts where the proportion of humus is minute, but the quantity of minerals is large. Plants, it seems, can grow anywhere; no matter how impoverished, imbalanced, or toxic a soil, plants will usually be found there, with leaves, root structure and chemistry tailored to fit.

Roots of desert plants typically spread widely and shallowly to benefit from dew (it has been said that, if turned root-side-up, a desert would appear fully inhabited), though some have long taproots which penetrate to deep underground water. Many desert trees do the same – tamarisk roots have been recorded (during the building of the Suez Canal) to the depth of 50 metres. In temperate dunes, grasses and sedges with long, creeping stems bind the sand. On tropical mudflats, mangroves grow stilt roots for stability.

In cacti and many other succulents, leaves have vanished, to be replaced by

spines. The growing of spines, thorns or prickles is a common response of trees and shrubs in arid places – savannahs, Mediterranean low scrub, and Asiatic steppe. Loss or reduction of leaves curtails water loss by transpiration; spines also act as a measure of insulation against the sun, help to collect dew, and discourage grazing animals. Plants of salty dunes and pebble beaches, by contrast, may be fleshy, like the aptly named saltwort *(Salsola)*, and the sea rocket *(Cakile maritima)*. Such plants that have adapted to very difficult substrates – deserts, salt lands, dunes or coasts – are obviously of great value: their guardianship of fragile lands is vital (pp. 36–7). Equally fragile, however, are the soils of the tropical rainforest, for all their lush greenery and year-round growth.

In northern Europe, and in much of North America, most of the lowland forests have long been cut down. In their place are rich productive farmlands and, where prairie agriculture has not taken over, a pleasant and sustainable landscape. In contrast, however, when tropical forest is cleared for planting with crops, disaster often ultimately ensues. The plants may flourish at first, but eventually the soil may be depleted, and torrents of rain wash away the thin soil. A tropical forest without its trees is not a lush paradise but can be turned into an unproductive and hostile place.

Of course, this stark contrast is not universal; but it is a common enough experience to create a great political dilemma. What right, the policy-maker in South America may say, have western experts to advise him not to fell his forests? After all, they have felled theirs over the centuries and are now

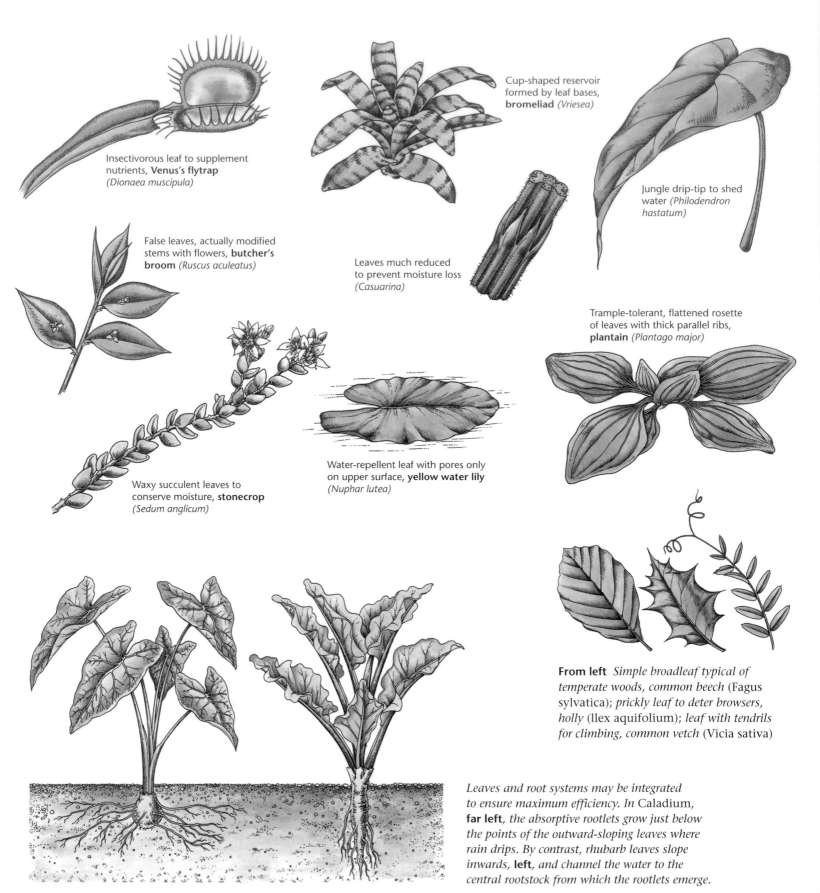

Insectivorous leaf to supplement nutrients, **Venus's flytrap** (*Dionaea muscipula*)

Cup-shaped reservoir formed by leaf bases, **bromeliad** (*Vriesea*)

Jungle drip-tip to shed water (*Philodendron hastatum*)

False leaves, actually modified stems with flowers, **butcher's broom** (*Ruscus aculeatus*)

Leaves much reduced to prevent moisture loss (*Casuarina*)

Trample-tolerant, flattened rosette of leaves with thick parallel ribs, **plantain** (*Plantago major*)

Waxy succulent leaves to conserve moisture, **stonecrop** (*Sedum anglicum*)

Water-repellent leaf with pores only on upper surface, **yellow water lily** (*Nuphar lutea*)

From left *Simple broadleaf typical of temperate woods, common beech* (Fagus sylvatica); *prickly leaf to deter browsers, holly* (llex aquifolium); *leaf with tendrils for climbing, common vetch* (Vicia sativa)

Leaves and root systems may be integrated to ensure maximum efficiency. In Caladium, **far left**, *the absorptive rootlets grow just below the points of the outward-sloping leaves where rain drips. By contrast, rhubarb leaves slope inwards,* **left**, *and channel the water to the central rootstock from which the rootlets emerge.*

In the tropical rainforest the layer of nutritious organic matter produced by leaf litter is extremely shallow, for its decomposition is hastened in the hot humid conditions. The weathered rock below is devoid of mineral nutrients, so roots do not penetrate far.

happy and prosperous. Western concern over deforestation can seem like a move to retain economic dominance. The consequences of this dilemma may be political, but its causes are biological.

Forests and their soils

The typical forest in a temperate country is usually on deep soil, which allows the tree roots to penetrate far down. There is plenty of leaf litter which forms the humus. Under the newest, dry leaves there is a deep, fine nutritious layer that readily supports small plants. Under the litter the soil is rich in minerals, left by the ice ages: either glaciers brought plenty of newly weathered particles, or old soils became well mixed with rock particles as ice froze and thawed within it. In tropical rainforest, by contrast, trees do not need deep or extensive roots – they get ample nutrition from the decomposing leaf litter on the ground, and sufficient moisture through their leaves. Beneath the litter is often found sand or clay. Unweathered rocks usually exist only at considerable depth, and most mineral content has been leached away.

Although leaf fall is virtually continuous, the layer of decomposing matter is shallow – heat and moisture accelerate the process of decomposition by fungi, microscopic animals and termites. The trees glean all the nutrients they can from the layer of decomposing litter, very little from further down, and hold the nutrients within themselves. Every bit of litter which drops is almost immediately sucked up again by this vast tropical vacuum cleaner. In effect the

tropical forest is a closed, nearly leak-proof system, millions of years old. Almost all the phosphorus, for example, is held in the vegetation rather than the soil, compared with half and half in a temperate oak wood; a good third of the nitrogen, as against a tiny fraction; and nearly half the magnesium, as against a sixth. Burning or felling the trees thus removes far more of the nutrients in the ecosystem than it does in a temperate forest.

In fact the fertile soil layer itself is so shallow that, without the trees, tropical downpours may remove it entirely, within one or two seasons. Gullies and serious erosion follow. The weathered rock particles, once exposed, can become like cement, forming an impervious layer of hard rock, called laterite. The farmer loses his crop; the forest

may not be able to regenerate, especially if large areas have been cleared. Even if it does, it will not contain the species diversity of the original forest. This, greatly simplified, is the tragedy of tropical deforestation.

Climate change and global warming

The related topics of climate change and global warming have seldom been out of the news in recent years. Most scientists now agree that human-induced rapid climate change is a reality, and that the influence of this on global weather patterns may be far-reaching, with major effects on our environment.

Throughout geological history the world has experienced cycles of natural climate change, with repeated periods of warmer and colder conditions, the extremely cold periods typified by ice ages. However, what we are seeing now is another effect, super-imposed on these natural cycles, and one that is entirely due to the activities of our own species – primarily caused by an increasingly rapid build-up of greenhouse gases over the last couple of hundred years. These gases come mainly from burning fossil fuels, and the main culprit is carbon dioxide.

The "greenhouse effect" itself is a perfectly natural phenomenon in which the atmosphere traps some of the warmth of the sun which radiates out from the earth's surface. If there were no shield of gas, acting rather like the glass of a greenhouse, the earth would be too cold to support life at all. But our own activities are adding to this effect and upsetting the natural balance, leading to a steady and rapid increase in the average global temperature (see graph opposite).

The damaging effects are likely to include more frequent droughts and increased desertification, and also less predictable weather patterns including possibly more frequent floods, and perhaps disruption of sea currents such as the warm Gulf Stream in the North Atlantic inducing paradoxical cooling to, for example, north-west Europe. As average temperatures rise, the ice caps begin to melt, and glaciers retreat, with consequent rises in sea levels. The latter effect in particular would be very dangerous for low-lying countries and oceanic islands, and increase the damage and loss of life caused by storms and tsunamis. By 2100 the prediction is that

global sea levels may have risen by between 9 cm and 88 cm, and average temperatures by between 1.5°C and 5.5°C.

Low-lying countries such as many oceanic islands could disappear – indeed this has already happened to some coral islands, such as in the Maldives group. Floods in low coastal regions would become more frequent. Moist tropical forests could become victim to natural forest fires as the climate becomes drier, destroying valuable habitats and also increasing the release of carbon dioxide, thus adding to the effect. In recent years it may be no coincidence that there have been huge forest fires in the Amazon region and also in Borneo. Food production could collapse in many African countries and else-where as droughts increase in frequency. Several terrible diseases such as malaria, which already kills two million people each year, may spread to afflict a wider swathe of the earth, and there is some evidence that this is already happening. The seas are already turning more acid as they absorb more carbon dioxide – this helps to reduce the amount in the atmosphere, but is beginning to kill corals and reefs which have taken centuries to form.

Many experts now regard climate change as an ecological time bomb, steadily ticking, and already responsible for increased insta-bility and unpredictability of climates, the world over. Reflecting the seriousness of this threat, a special taskforce of scientists and politicians produced a rather chilling report about global warming (*Meeting the Climate Challenge*, January 2005). This report points to a figure of global warming, a danger

point in the temperature rise beyond which disastrous changes will be impossible to avoid. This figure is considered to be a rise in the average global temperature of 2°C above pre-industrial levels.

The conclusion is that warming must be kept below this in order to avoid catas-trophe. This in turn means controlling the levels of carbon dioxide (the main green-house gas) to below 400 parts per million. Yet this level already stands at 370 ppm, and is steadily rising. In the late 1950s, the average levels were about 315 ppm. The solution is easy to say – reduce emissions by 50 percent by 2050 – but less easy to achieve as governments pursue their own development and economic goals. But the fact that there is now general global agree-ment about the problem is a start to seeking ways to combat the threat which will surely affect us all.

Existing technology can help in the fight against global warming. Possible solutions are to cut energy waste, use more renew-able resources including solar and wind power, and bio-fuels, grow more trees (which tend to remove carbon dioxide from the air), capture gas released from power stations, and (more contentiously) create more nuclear ("clean") energy.

One of the difficulties is that carbon dioxide remains in the atmosphere for a long period – up to 100 years, so that there will be a considerable lag before cuts can take effect. But even modest cuts in green-house gas emissions should slow the rate at which the climate warms, giving ourselves and the planet's wildlife time to adapt to the changes that ensue.

Ice melt due to global warming, Spitzbergen, Arctic.

Recent findings based on measuring the temperature of the oceans and comparing these readings with data going back some 40 years have provided more evidence that global warming is a reality. The steady warming of the seas over this period cannot be explained by natural causes, and the only conclusion is that human-induced global warming is the culprit. Recent studies of ice sheets and glaciers point in the same direction. The world over, ice is in retreat – from the Arctic and Antarctic, to the high Andes, the Himalayas, to the Alps and mountains of Norway. In March 2005, the snow-capped summit of Mount Kilimajaro in northern Tanzania melted to reveal the tip of Africa's highest peak for the first time in 11,000 years.

The glaciers of the Himalayas are melting rapidly, some retreating at more than 15

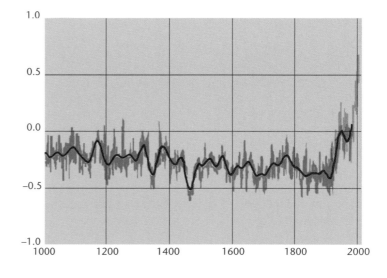

Northern hemisphere temperature

This graph is a reconstruction of global temperature fluctuations since AD 1000. Note the dramatic increase in temperature in the last 50 years.

metres each year. The initial effects of this are to increase the volume of water in the major rivers, causing more frequent floods, but after a few decades, the water levels are expected to drop, causing environmental problems for hundreds of millions of people in northern India, Nepal and western China.

By working together and applying new technologies to combat the problem we have the opportunity and also the tools to slow and possibly halt this runaway warming process. If we succeed, it will certainly be to the benefit of human life and also of the natural ecosystems on which we all depend.

Guardians of diversity

The rainforest vegetation found throughout the tropics is amazingly complex. For one thing, there is usually no particular plant that predominates. A hectare and a half of Malayan forest, for example, has been found to contain 227 kinds of trees – but there might be only one or two individuals of a species. One study of the Brazilian rainforest found a staggering 487 tree species on a single hectare. A similar area in deciduous woodland in Michigan, USA could contain as few as 10 species, but many individuals of each: in coniferous forest near the Arctic Circle, it might contain one tree species only.

A second level of complexity derives from the vast quantities of plants and the diversity of ways in which they exist. For support, the trees may have stilt roots in wigwam arrangement; curiously enough many of these lose the original central stem of the seedling tree when they grow to any size. The alternative method of support is the

buttress root. If one sees an uprooted buttressed tree, it is astonishing to observe how its feeding roots jut down from these buttresses for as little as 50 cm. Many other trees have no such architecture and are supported by each other. If a clearing is cut through the forest, the ordinary thin-trunked trees sometimes just fall over into it: their roots have only a weak anchoring capacity.

The rainforest contains some very large trees, maybe 50 metres tall, and varying layers, or storeys, not always clearly defined, of lesser trees below. The tallest, overtopping trees are called emergents, and between them the space is filled by the wide crowns of the main canopy trees. A layer of smaller trees, or young individuals of the taller species, is typically found at around a quarter of the height of the emergents. There is then a shrub layer – some seedling trees, but mostly mature low-growing species adapted to the reduced light; around them will be herbaceous plants of many kinds.

Within this deep forest the level of air humidity increases fourfold from top to floor; but wind, rainfall, and temperature (and temperature fluctuation) are at their lowest on the floor, and greatest above the canopy.

Climbing plants germinate on the floor and ascend the trees in various ways. Thin-stemmed species like philodendrons attach themselves to trunks with aerial roots (just like ivy in temperate climates). Once into the canopy, they send down vertical aerial roots which absorb moisture from the air and finally anchor themselves in the ground. The lianas (like temperate clematis) have thick stems, sometimes round but more often

Most rainforest trees have such shallow root systems that they effectively support each other. Some have special ways of ensuring stability: they may have stilt roots produced from the lower part of the trunk, as shown **above***, or thin, wide-spreading flange-like buttresses,* **below***. These dramatic buttresses send down thin vertical feeding roots, only 50–100 cm long, into the shallow soil layer.*

flattened and twisted, or pierced and ladder-like; they twine round trunks and branches. In the Far East, there are rattans or climbing palms with downward-pointing spines to prevent the thin, upward-thrusting stems from slipping down; they are like the giant clambering roses of the Himalayas.

The converse of the climber is the strangler. These are trees – the familiar house-plant *Ficus benjamina* is one – which will grow in debris in another tree crown, if their seeds have germinated there, and send down roots that eventually reach the ground. Criss-crossing round the trunk of the host tree, these roots coalesce and thicken, and literally prevent the host from expanding its trunk, while the strangler's leaves crowd out those of the host. After many years, the host dies and decays, to leave just the woody cylinder of the strangler's roots and stems.

On the spreading upper branches of the canopy trees especially, there will be a dense growth of ferns, orchids, bromeliads and other epiphytic plants, content with a limited amount of moisture absorbed from the air and in debris which accumulates around their anchoring aerial roots. Some grow lower down, notably great ferns like the "birdsnests" and "stagshorns".

Lichens, mosses and algae colonize all levels of the forest, treating tree trunk and rock the same. And everywhere on the floor, growing on fallen leaves and tree trunks, there will be fungi in an amazing array of form and colour. They serve to break the dead matter back into its constituents, to feed both the old inhabitants of the forest and equally the young seedlings emerging and hoping for a break (in both senses of the word) so that they can reach the desired light level without being choked.

Guardians of the water cycle

People have cut down the trees around them for a very long time. The civilizations of Mesopotamia and the Indus Valley foundered partly because of excessive defor-estation. In Greece, in the fourth century BC Plato wrote that "Compared with what it was, our land is like the skeleton of a body wasted by disease. The plump soft parts have vanished, and all that remains is the bare carcase". Whatever the reasons behind the present accelerating rate of forest clear-ance, the removal of trees usually demon-strates how essential they are – or have been – as guardians of the environment. Trees safeguard the soil on the steep slopes of mountain watersheds, where rivers have their sources. Cut down the trees and the soil is washed away off the underlying rock, sometimes gouging deep ravines. Landslides occur. The water rushes into rivers which cannot contain the sudden onslaught and overspill their banks downstream. When the floods eventually subside, there can be long periods when the rivers are dry, for the slow-releasing sponge of the forests is no longer there to trickle moisture into the streams.

In China, disastrous floods in 1981 killed thousands of people and made one and a half million homeless. Deforestation of the upper reaches of the Yangtze River was publicly acknowledged as the cause of the tragedy. There were also serious floods in

Kapok (Ceiba pentandra) *is an important component of both Old and New World tropical forests, especially in areas where there is a dry period; the forest type (in contrast to that of the savannah) is a large tree with buttress roots. The seeds are surrounded by very light fibres, at one time of considerable commercial importance but now partly replaced by synthetics.*

Epiphytes of the tropical rainforest include many orchids like the Brazilian Maxillaria picta, **right**. They cling to tree branches with aerial roots and absorb sufficient moisture through these from the air, and debris which accumulates around them, which will include nutrient minerals. Water reserves are held in the fleshy "pseudobulbs" at the base of the leaves.

Bromeliads such as vrieseas, **far right** – relatives of the pineapple – are found on trees and cliffs in areas ranging from very hot to quite cold in much of South and Central America. They are mostly epiphytes with minimal roots for attachment, and a unique system for ensuring water supplies – the leaves form a central reservoir in which rain collects, with special water-absorbing cells at their bases.

A curiosity of tropical jungles is the strangler, **below**. A number of different trees are involved. Their seed germinates in the debris-filled crown of the "host"; as it grows, the strangler sends down roots which eventually reach the ground and thicken, often meshing together around the host's trunk. Eventually this prevents the latter expanding and may choke it to death.

Below Rainforest vegetation can occur at altitudes from sea level to mountains, as in this Andean example at about 3,000 metres; at such high altitudes it may be known as cloud forest. High air humidity is common to both, encouraging epiphytes and ferns as seen here, together with fungi which play an essential part in the forest's life cycle by decomposing dead vegetation into humus.

1998, especially in Hunan Province where nearly one-fifth of the rice harvest was lost, 2,000 people died and six million homes were washed away. In the summer of 2004, flash floods and landslides killed 500, caused 1.5 million people to flee their homes, and ruined five million hectares of farmland.

Trees intercept and hold water. Moist air is continuously produced over them, clouds form, rain regularly falls. Estimates for the proportion of water vapour recycled in rainforests range from 15 to 75 percent. What is becoming apparent is that without the forest, rainfall is reduced. At Mahbeleshwar, south of Mumbai, where the surrounding hills have been severely logged, annual rainfall has decreased from 1,000 cm a year to a little over 600 cm, and the pleasantly warm summers have become oppressively hot. When virgin forest was originally cut in Malaysia to plant rubber, rain clouds were seen to stop precisely at the edge of the tall forest and left the young rubber plantations dry.

The rainforests occupy a large part of the tropics (30 percent of the land area of the 20° equatorial belt lies in the Amazon region, for instance), and the air masses associated with them are globally very important. Upsetting their hydrological balance is likely to cause worldwide disturbance to weather systems.

That the world's forests are in crisis is indisputable. About half of the original forest cover has been lost, and only about a tenth of what remains is protected, and most of this is badly managed. However, it is not all bad news. The Forest Stewardship Council (FSC), works with WWF to encourage protection of natural forests and sustainable use of managed forests. Encouraged by this support, Brazil now has over 2,300,000 hectares of certified native forest and plantations, and Indonesia has recently established nine new national parks, which protect forests that are home to many species, including Sumatran tiger, elephant, and orang-utan.

India has also shown the world a way of combating deforestation. In the 1970s, locals in Uttar Pradesh resisted forest destruction through what became known as the "Chipko" movement. When contractors arrived to cut timber, women from the villages, following an age-old precedent, embraced the trees, hoping to save them from the axe. This grassroots movement, called "Chipko" ("embrace"), proved extremely effective in protecting certain Himalayan forests. It is mainly a women's movement as it is the women who suffer most from having to go further afield to graze their cattle, gather firewood and find wild fruits, nuts and honey. Now the women are starting to replant some of the forests, including the many wild plants that they use in daily life. Chipko also influences government: in 1980, Chipko leaders persuaded Mrs Gandhi to ban felling on steep Himalayan slopes and the movement spread to other parts of India.

Trees on mountain watersheds safeguard the soil on the steep slopes by binding it with their roots, and absorbing much of the rainfall. If the trees are cut, or killed by excessive tapping for resin, this guardianship of the terrain is entirely lost. Soil is washed away, rain gouges out ravines, exposing the underlying rock, causing landslides, and creating all the horrors of erosion in valleys lower down, as in this all too typical scene, **above**, *in the highlands of Nepal.*

Loss of India's forests is now a major political issue. Floods caused by deforestation in the Himalayas mean that an estimated 6,000 million tonnes of soil containing a million tonnes of nutrients are lost every year. The 1979 monsoon did over 2 billion dollars worth of damage: in Bihar, West Bengal and Uttar Pradesh 3,000 people and at least 200,000 cattle were killed by floods, and the following drought destroyed 35 million hectares of cropland.

In 2001, dreadful floods, brought on by a prolonged monsoon, affected 8 million people, destroying 60 percent of the coastal rice crop in the state of Orissa. In 2003, tens of thousands of people were affected by landslides and flash floods in West Bengal, with the tea-producing district of Darjeeling badly hit.

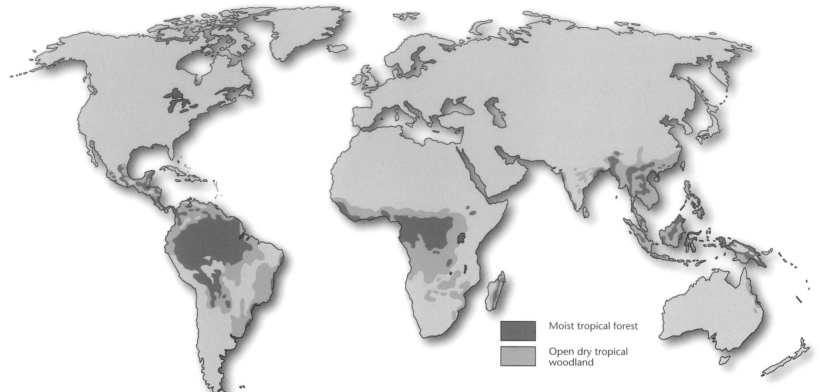

Moist tropical forest

Open dry tropical woodland

Grazing animals in excessive numbers are a prime cause of almost irremediable damage and desertification: in the Mediterranean the goat is the major culprit, especially as it can readily climb into trees like this carob. Cattle and camels are the other main culprits.

Chipko leader, Sunderlal Bahaguna, has said, "In one of our religious books it is written that a tree is like ten sons. It gives ten valuable things: oxygen, water, energy, food, clothes, timber, medicinal herbs, fodder, flowers and shade."

Tropical forests

*About half the world's original forest cover has been lost, and only about a tenth of what remains is protected. The map **above** shows the extent of the moist tropical forests, which are particularly at risk, and areas of mainly dry, open woodland, some of which results from degradation of rainforest. The losses of natural forests in temperate regions, especially in Europe and North America, have been more dramatic, with only about one percent of the original cover left in these regions.*

According to a UN Food and Agriculture Organization report (2001), nearly 37 million sq. km of natural forest remained in 2000, and 1.9 million sq. km of plantations. One of the largest of these expanses was formed by the Amazon rainforests.

The net loss of natural forest in the 1990s was 1.2 million sq. km, or 3 percent. Although Brazil had the greatest loss in terms of size, this was only a 4 percent change. The countries that lost most in proportion were Burundi (60 percent) and Haiti (57 percent).

Brazil's Atlantic rainforest now covers no more than 5 percent of its former area, and is the subject of major international rescue efforts. Most of the lowland forests outside protected areas of peninsular Malaysia and the Philippines have been logged.

Guardians of fragile lands

The world perhaps first woke up to the devastating consequences of destroying the natural guardians of the soil, when the dust bowls of the American Middle West were created in the 1930s, with results on land and people so poignantly described in John Steinbeck's novel *The Grapes of Wrath*. These were areas where the natural vegetation had been removed for growing crops, mostly cereals, with the addition of fertilizer but no humus-forming matter. After a relatively short period of heavy cropping and annual ploughing, the soil of the arable prairies simply crumbled to dust and blew away.

Agricultural land commonly has soil erosion rates at least five times the natural level; removal of hedges and trees, and lowering of water tables (as in fenlands) make matters worse. Even Europe, the continent least troubled by soil erosion, is calculated to be losing close to a billion tonnes a year. In parts of the world where forest cutting and pressures on the land are severe, the rates are far higher – one estimate suggests that human disturbance may have increased the global denudation rate from 20 million to over 50 *billion* tonnes a year.

Every year, more than 100,000 sq. km of cropland are degraded and lost through erosion by wind and water, and 4.3 million sq. km of arable land have been abandoned in the last four decades alone.

The most vulnerable soil of all is that of the world's drylands – areas where rainfall is low and the climate usually harsh. Covering about a third of the earth's land surface, their natural vegetation includes grassland,

low woodland of scattered trees with grass, and steppe. Over-grazing and cutting of the trees for firewood gradually removes the plant cover. As the plants are eaten away, the climate grows drier, while the pressures from hungry herds of livestock intensify – a vicious circle that ends in barren lands. The livelihoods of hundreds of millions of people are threatened. Drylands used for agriculture total some 52 million sq. km, but nearly 70 percent of this land is degraded or affected by desertification. In Africa and in North America nearly 75 percent of agricultural drylands are already seriously degraded. Over the past 50 years, an area the size of India and China combined (12 million sq. km) has been subject to moderate to extreme soil deterioration.

The combating of erosion and degradation of lands that once provided a livelihood is clearly one of the most urgent tasks facing humanity – and one in which plants, particularly those tailored by nature to survive in fragile lands, play a crucial role.

Hardy acacia species, for example, planted in the path of shifting dunes, are helping to halt the desert's march. Trees in desert conditions have a hard time, but if they can be established they both bind the sand and gradually accumulate soil, resulting from the decay of falling leaves around their trunks. Stockproof fences are erected and the sand stabilized by "planting" rows of dead brushwood, after which the acacias and other trees are planted closely and provided with enough water for establishment. This method resembles those being tried in British peatlands, where erosion is being halted by planting

rows of straw, or by drilling seed into the unploughed stubble of a previous crop.

There are other possibilities. Alginates (product of seaweeds) and granular synthetic polymers are substances that will absorb many times their own weight of water – one polymer now in experimental use absorbs up to 30 times as much. When these are worked into the upper layer of sand and watered, annual weeds usually sprout. These can either be ploughed into the soil to start the humus cycle, or crops or trees can be planted directly.

It is amazing what can be achieved by tree planting. The Chinese are perhaps the greatest present day exponents; literally millions of trees have been planted in the last few decades. Even in the stony wastes of the Gobi Desert, plantings of poplars amaze the visitor.

With modern technology, crops can be conjured from the most barren lands. With irrigation, Saudi Arabia has gradually increased wheat production to the point where it is now a net exporter. Deep beneath the arid crust lies a hidden treasure of aquifers stretching eastward from the Arabian Shield to the Gulf. But the water is a fossil resource; it fell 15,000 to 30,000 years ago, and though it is calculated to last at least 100 years, like oil it cannot be renewed.

Israel has a different approach to desert agriculture. Essential low-value commodities such as grain and sugar are imported, and the desert is used as a huge natural hothouse providing goods such as fruits, vegetables and cut flowers for luxury markets. Israel, too, draws on underground

*This astonishing photograph of the edge of
the Gibson Desert in western Australia shows
trees maintaining individual patches of fertile
ground around them: their leaf litter produces
organic matter in which soil organisms can
multiply and humus is created, thus retaining
the very limited amounts of moisture
available for continued root growth.*

water sources, but has an advanced
system of management, with less than two
percent of all run-off allowed to flow to the
sea unused.

Comparable aquifers and high water
tables exist in many countries, but where
they are being tapped they are falling at a
more or less desperate rate. In the American
south-west, arid-land agriculture is clearly
beginning to disintegrate, largely as a result
of groundwater depletion in the first quarter
of the twentieth century. Large areas of
riparian forest and mesquite woodland
died as a result.

Allocating water resources in the western
United States has now become an issue of
contention between California and other
arid western states. The argument centres
on the question, how much water can each
state, with its city-dwellers and farmers, take
from the great rivers, such as the Colorado,
which flow from the Rockies to the Pacific?
With limited resources and virtually unlimited demand, there is no easy answer.

One way and another, we may be able to
overcome these problems, but at what cost?
Is it really worth the huge input of energy
and technology to force the deserts to
bloom unnaturally with carnations and
cabbages? Some believe the way forward is
to adopt new crops using plants native to
arid lands, such as jojoba *(Simmondsia
chinensis)* of northern Mexico and southern
USA. Jojoba produces a high-quality oil and
grows well in poor soils.

Super-salt-resistant plants may also be
valuable: large areas of formerly agricultural
land cannot now grow crops because of a
build-up of salt following short-lived

The Great Salt Lake Desert in Utah, USA is a land encrusted with salt crystals, dramatic but lifeless. Little can be done with deserts as salt as this one, but potential crop plants are being developed for saline situations – often the result of faulty irrigation techniques – so that terrain once thought totally unproductive can some day be made productive.

irrigation schemes; other areas are naturally too salty for normal crops. But for example, the leguminous tamarugo tree *(Prosopis tamarugo)* from Chile can survive in exceptionally salty conditions and (according to the US National Academy of Sciences) its highly nutritious pods and leaves allow sheep to be stocked at rates close to those of good grassland.

The deserts are ripe for the new wonder crops, but there is still plenty of scope for traditional methods. Savannah and desert scrub have been cultivated for centuries without ecological failure, and it does not take new crops to end the profligate use of water. And the social fabric of a region will be much less disrupted by modifying old technology.

For example, the Sudanese have adapted a natural ecosystem with *Acacia*-grassland as the cultivation climax, to produce gum arabic as their second largest export. Native farmers clear the scrub, and the cleared land is planted with millet, sorghum and sesame until after a few years the poor soil is exhausted and the land is left fallow. Among the native flora that regenerates and restores fertility is *Acacia senegal*. This is tapped for gum arabic in the latter half of the soil-resting period.

Unfortunately this ecologically balanced cycle has broken down in most places because under the pressure of a growing population the cultivation period has been over-extended, severe grazing is preventing seedling establishment, and mature *Acacia* bushes are lopped for firewood. However, the system has been further modified: *Acacia senegal* has been included in shelter belts

Subterranean aquifers may contain huge amounts of water but they are non-renewable resources. Pumping them out is usually of short-term value, made ridiculous in situations where hot climates quickly evaporate the water. In Jordan, run-off from irrigation systems is returned to the aquifers together with surplus winter rain, but chemicals are beginning to accumulate in the subterranean reserves with potentially alarming results.

Rhizophora

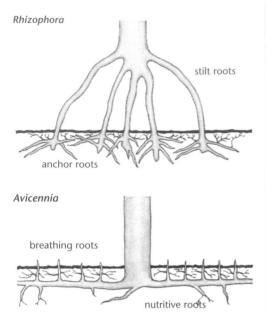

stilt roots

anchor roots

Avicennia

breathing roots

nutritive roots

Mangrove, or tidal forest, is the natural climax vegetation of estuaries and sheltered muddy inlets in the tropics. Several distinct genera have adopted similar lifestyles to live in these muddy tidal places. Many like Avicennia (lower picture) have "breathing roots" or pneumatophores which are specialized mechanisms for dealing with the airless mud: their pores allow gases but not water to pass through. When the tide is out they absorb oxygen for a fast bout of respiration, providing energy for the controlled absorption of nutrients from the mud by the feeding roots below. Other mangroves, like Rhizophora (upper picture) have stilt roots which are primarily for support in the shifting, very fluid mudlands. The mud accumulates around the roots and slowly builds to soil. The intricate network of roots and stilts protects soft-soiled coasts from wave erosion and flooding. Mangroves produce fuel and timber and harbour many kinds of fish and shellfish, especially shrimp larvae.

Pollution, rubbish dumping and other uses of tropical tidal forests disrupt or destroy these fragile ecosystems, threatening the livelihoods of millions of people in developing countries who are directly or indirectly dependent on them.

Mangrove forest, Sundarbans, Bangladesh

Right and below *Marram grass* (Ammophila arenaria) *is common on sand dunes out of range of the tide throughout western Europe. The root system is deep and wide spreading, and its long creeping stems root down at the nodes. Marram-stabilized dunes can reach over 30 metres high, and grow 30 metres laterally and 8 metres vertically over a 10-year period.*

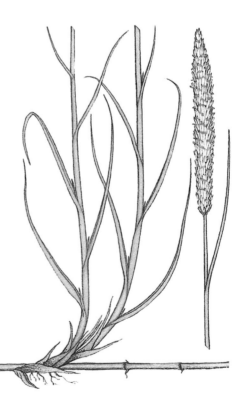

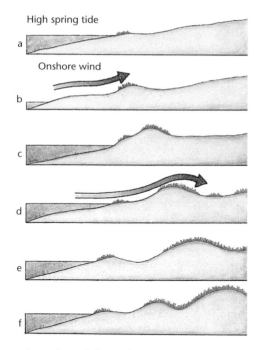

High spring tide

a

Onshore wind

b

c

d

e

f

Coastal sand dunes form and stabilize through the agency of plants. Pioneer species such as Atriplex littoralis *germinate on the drift line at the upper reach of spring tides (a). Wind-borne sand collects around them and an embryo dune forms (b). New species such as marram grass* (Ammophila arenaria) *invade. As more sand is deposited the plants push up new shoots to counteract burial (c). As the sand accumulates on the seaward side, the marram grass spreads through it, thus pushing the dune seawards. At the same time the dune grows landwards as the onshore breeze carries sand up the windward face and deposits it in the lee (d). Vegetation binds the sand and covers the dune with a protective mat. But if this blanket is broken, by regular trampling from vacationers for example, the sand can easily blow away. More and more species colonize the dunes (e). As sand accretion lessens, the* Ammophila *is replaced by other species and eventually mosses and lichens become prominent. Different species are found in the depressions between the dunes which are often damper (f). Eventually the old dunes start to flatten, while a series of younger dunes forms to windward.*

around towns, where people are licensed to collect gum.

Plants which, like *Acacia*, are adapted to difficult conditions, offer great hope for land reclamation. Many plants are superbly adapted to the sites where they grow wild. Desert species often have mechanisms for searching out, storing and conserving water. Those on seashores and dunes may be adapted to similar problems of physiological drought, because the strongly saline conditions prevent the roots absorbing water readily. Here too, plants act as guardians, like marram grass (*Ammophila arenaria*) stabilizing dunes along temperate coasts. Colonization of shores by the saltmarsh and sand dune plants finally enables enough soil to accumulate to bear more orthodox plants. Glasswort (*Salicornia*) and cord grass (*Spartina*), for example, garner particles on the retreating tides, and so slowly build marsh from open mud. For many centuries such new land has been reclaimed for agricultural use.

The most remarkable guardians of the coast are the mangroves, which colonize mudflats near the Equator. For stability in the slushy mud they grow in, many develop stilt or prop roots. But because the mud is without the oxygen that is normally found dissolved in water (which enables most aquatic plants to prosper), these plants produce finger-like or looped breathing roots. Their role in protecting low-lying, sandy or soft-soiled coasts against the destructive pounding of the sea is vital. Recent damage by tsunamis in the Indian Ocean would certainly have been less extreme if more of the natural mangrove vegetation had been conserved.

Even the squalid deserts of modern industrial society – rubbish tips and land spoiled by pollution – can be reclaimed by plants. In the past, the only certain way of getting good cover on waste heaps was to top them with an expensive layer of inert material so that standard seed mixtures could be used. By trial and error a number of plants that grow in toxic sites have been found. Some are grasses, such as brown bent (*Agrostis canina*); there are members of the pea family like clover and gorse; and trees such as alder and silver birch. Better still is to find the plants that nature has crafted to deal with the problems; examples are the grasses rat-tail fescue (*Vulpia myuros*) for copper waste, and the derived Merlin variety of red fescue (*Festuca rubra*) used on calcareous lead and zinc wastes. Even the humblest grasses thus have their role of protector and provider.

Every part of ecosystem earth depends on the green life-support apparatus for its survival as a whole. Most valuable of all is the forest: in the words of the Buddha: "The forest is a peculiar organism of unlimited kindness and benevolence that makes no demands for its sustenance and extends generously the products of its life activity: it affords protection to all beings, offering shade even to the axeman who destroys it."

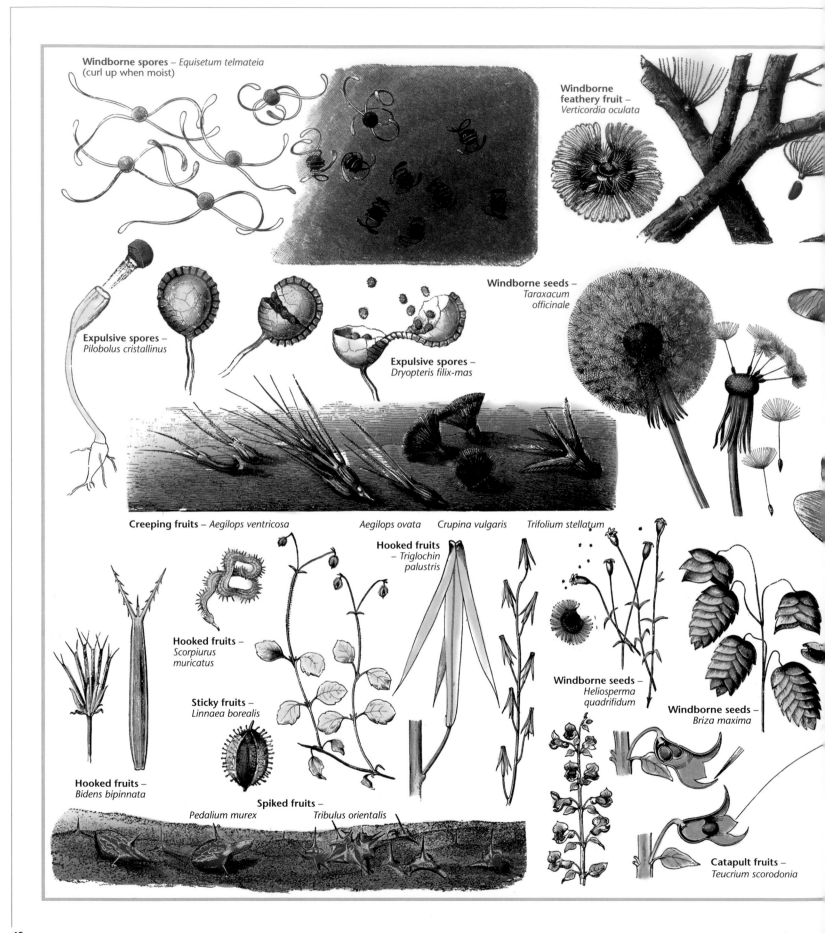

Windborne spores – *Equisetum telmateia*
(curl up when moist)

Windborne feathery fruit – *Verticordia oculata*

Expulsive spores – *Pilobolus cristallinus*

Expulsive spores – *Dryopteris filix-mas*

Windborne seeds – *Taraxacum officinale*

Creeping fruits – *Aegilops ventricosa*

Aegilops ovata

Crupina vulgaris

Trifolium stellatum

Hooked fruits – *Triglochin palustris*

Hooked fruits – *Scorpiurus muricatus*

Sticky fruits – *Linnaea borealis*

Hooked fruits – *Bidens bipinnata*

Spiked fruits – *Pedalium murex*

Tribulus orientalis

Windborne seeds – *Heliosperma quadrifidum*

Windborne seeds – *Briza maxima*

Catapult fruits – *Teucrium scorodonia*

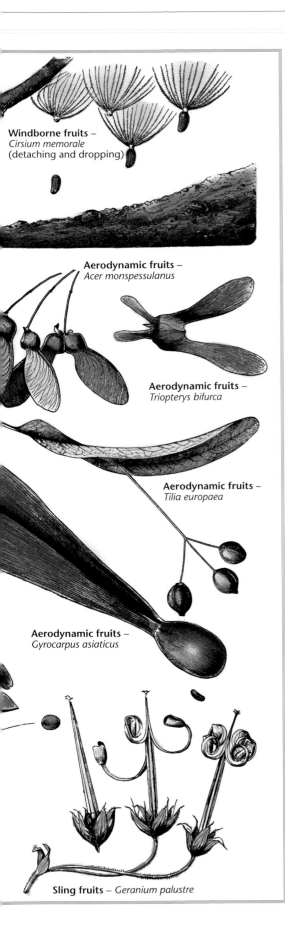

Windborne fruits –
Cirsium memorale
(detaching and dropping)

Aerodynamic fruits –
Acer monspessulanus

Aerodynamic fruits –
Triopterys bifurca

Aerodynamic fruits –
Tilia europaea

Aerodynamic fruits –
Gyrocarpus asiaticus

Sling fruits – *Geranium palustre*

CHAPTER THREE

Green Travellers

Plants are very easily distinguished from animals – the most obvious difference being, of course, that plants are generally static, whereas most animals move around. Yet as every gardener knows this immobility of plants does not seem to be much of an impediment – they will arrive rapidly enough on any bare patch of soil! Plants have spread across the globe every bit as effectively as animals; indeed they managed to establish themselves on dry land well before animal life could do so, over 400 million years ago.

The success of plants as colonizers depends on a remarkable characteristic: that small portions of a plant can survive away from the parent, to be moved by wind or water, insect, bird or mammal. Seeds and spores, of course, are the best known and most abundant of these "portions" – if they settle where conditions are favourable, they can grow into whole new plants. But many plants can also spread vegetatively. Their means of dispersal are varied and surprising; among the most remarkable cases, perhaps, being the "tumbleweeds" of the steppes – the whole plant becomes detached and is blown around by the wind, scattering the ripe seeds.

Many plants have adapted for dispersal by animals, and may have fruits or seeds which cling to fur, or to shoes and clothing. With the arrival of humans, a new method of plant movement was added. At first, travellers carried plant material as food on their journeys, and discarded, dropped or deposited the seeds away from their native homes; later they carried many of the same seeds, plants and even trees around the

Wandering continents

Magnolia campbellii

world with the deliberate intention of extending their natural range. Modern world plant distribution is massively affected by humans, and many apparently "natural" landscapes are, in fact, largely a result of human influence: the autumnal colours of New England woods are partly due to introduced species; suburban streets are lined with cherries from Japan or ginkgo from China; crops are grown on alien continents; gardeners can take their pick of the world's flowering species. The rate of transfer is increasing – seeds and spores, for example, can now "hitchhike" on jets and liners, to spread around the world. Human-induced movement of plants is not necessarily a good thing. Just as overenthusiasm for collecting plants threatens some habitats, the introduction of exotics can damage others, swamping the native flora and its dependent animal communities. Tragically, many thousands of plant species are now in danger of extinction through human irresponsibility.

Plant origins

A hundred years ago, only a few eccentrics believed that the continents moved. Today, however, the concept of continental drift is central to earth science. The evidence points to the existence, about 225 million years ago, of a single supercontinent – Pangea, or "all-earth". This began to break up about 200 million years ago, forming the large landmasses of Laurasia to the north, and Gondwanaland to the south. The earliest land plants date from long before this, and the extraordinary history of our moving

continents underlies the present day distribution of many plants. Roughly 390 million years ago such primitive forms as liverworts, mosses, and horsetails had emerged, to be followed shortly by ferns. Conifers began to appear around 225 million years ago, and spread over much of the land surface. Charles Darwin called the origin of flowering plants "an abominable mystery" – the fossil record is almost blank as to how the evolutionary leap from ferns and conifers to true flowering species took place. Certainly they must have evolved and spread very rapidly indeed, on a geological timescale. The earliest fossil pollen and also fossil seeds date from the Upper Devonian period, about 350 million years ago; most is later. Yet present day distribution suggests that flowering plants, or their immediate ancestors, must have colonized Pangea before it had completely broken up, when dispersal was possible both within Laurasia and Gondwanaland, and also between them, via landbridges which still existed some 200 million years ago.

While rhododendrons show their Laurasian origin by a basic distribution in North America, Europe, Eurasia and Southeast Asia, the heaths *(Erica)*, in the same family as rhododendrons, have their centre in South Africa with a few species in the Mediterranean and Europe.

The fact that seeds get moved around does not necessarily mean that plants can establish themselves everywhere. There are both physical and climatic obstacles. High mountain ranges are usually effective barriers: many plants found west of the Rocky Mountains of North America are not

Giant or king protea
Protea cynaroides

Erica doliiformis   *Erica infundibuliformis*

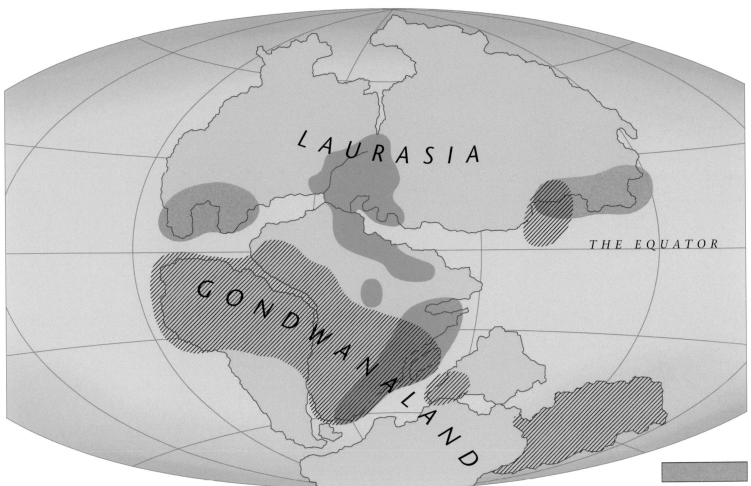

200m. years ago

About 225 million years ago there was a
single landmass, known as "Pangea"
(meaning "all-earth"). By about 200 million
years ago this was beginning to break up, and
comprised Laurasia (now North America,
Europe and Asia) and Gondwanaland (South
America, Africa, Antarctica, India and
Australia). At this time flowering plants were
evolving and spreading: some current
distributions are shown, **above**, overlain on
the ancient continents. Magnolias, primitive
in floral terms, show a largely Laurasian
distribution: eastern USA, Mexico, Japan,
China and the Himalayas. A spectacular
example of Gondwanaland spread is shown
by the protea family. Their multiflowered,
brightly coloured heads vary from the
bottlebrush shapes of Australian banksias to
the showy bracts of South Africa's king
protea, **middle left**. The ericas evolved after
the breakup of Pangea; their distribution
reflects more recent continental evolution.

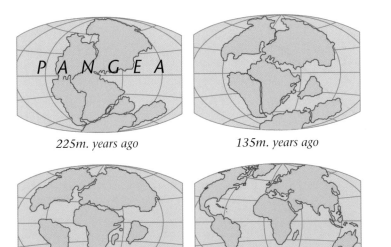

225m. years ago

135m. years ago

65m. years ago

Present day

Magnoliaceae

Proteaceae

Erica

43

*Krakatoa provides a perfect example of tropical succession. In 1883 Krakatoa, **above**, six months before the explosion) exploded leaving Anak Krakatoa completely devoid of vegetation.*

*Botanical expeditions have charted the recolonization: the island was first dominated by algae, then by ferns, then by flowering plants and finally by trees, so that now part is clothed in dense jungle, **right**. More than 100 other plants did become established, only to disappear in the interim. The diversity of the flora continues to increase, though it is still not as rich as the forests on nearby Java and Sumatra.*

The cactus Brachycereus nesioticus, **above**, *is restricted to the Galapagos Islands where it grows only in certain coastal regions on hard black lava flows. Lodging in crevices on the soil-less rock, it makes wide-spreading clusters where virtually no other plant can survive.*

found east of the range and vice versa; *Magnolia*, for example, is found east of the Rockies (and globally east to China) – but not to the west.

Climates worldwide have also changed considerably during measurable geological time and plants seem to have adapted to these changes relatively readily. Today many plants, adapted to one kind of climate, cannot transfer from one major climatic zone to another: plants from the tropics will not thrive outdoors in temperate zones (though most of the west's popular house-plants are tropical), nor vice versa. Temperature, seasonal patterns and day length (which can affect flower and hence seed production) are among the main controls. These problems are very plain to gardeners especially in North America where plants are designated to no fewer than ten zones according to what minimum temperature they will stand. Even in Britain, what can be cultivated in the Scilly Isles differs vastly from the possibilities of the Midlands, about 300 miles north-east.

In comparatively recent times ice has been one of the the great controllers of plant life in what is now the temperate zone (parts of which were once tropical, including southern Britain and much of the rest of Europe). There have been several ice ages,

the most recent one reaching its peak some 20,000 years ago. Ice sheets spread southwards from the Arctic into North America, Asia and Europe, obliterating life and cooling the climate. About 14,000 years ago the global climate warmed and became moister, probably over just a few decades; the ice began to retreat and the forests to spread again, only to be knocked back by another colder phase which lasted about 1,300 years, after which the climate reached roughly the conditions we experience today. These facts help explain why relatively few plants are found in northern Europe – they have had only about 10,000 years to re-establish themselves. The present day flora is also adapted to the now cooler climate.

The effects of the ice ages were not limited to the temperate zone; although temperature fluctuations were much smaller in the tropics, climate patterns were sometimes severely disrupted, particularly in Africa.

Colonization of virgin habitats is spectacularly rapid in warm climates. It is best observed on volcanoes where ferns can germinate on the surface within six weeks of an eruption while it is still hot. But the most dramatic example of recolonization and demonstration of how plants can travel is that of Krakatoa, the island in the Sunda strait between Java and Sumatra which

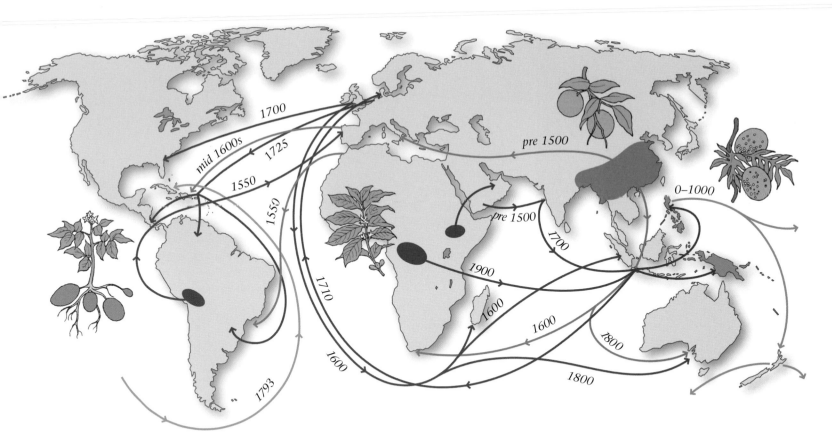

The spread of crop plants

The map shows how four crop plants spread around the world through human influence. Citrus (oranges etc.), following the overland Silk Road from China, travelled to Europe in ancient times. Coffee from Africa, potato from South America and breadfruit from Tahiti travelled across the oceans to distant continents. Coffee and potato have both had an extraordinary series of migrations, their spread largely related to the expansion of European colonial powers between the fifteenth and nineteenth centuries.

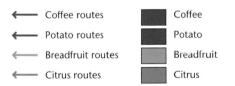

←——— Coffee routes
←——— Potato routes
←——— Breadfruit routes
←——— Citrus routes

■ Coffee
■ Potato
■ Breadfruit
■ Citrus

exploded most violently in 1883. After the cataclysm three islands of quite new outline were left, covered with ash and pumice 30 to 60 metres deep. As might have been expected the first arrivals were plants that required little or no soil and which were wind-borne over great distances. These included cyanobacteria, accompanied by lichens and mosses, and no fewer than 11 kinds of fern within the first three years. Soon there were also 17 kinds of flowering plant. Nine had arisen from seeds washed up on the beaches; of the remainder, four species were composites and two grasses, all with light, airborne seeds. Within 14 years the number of colonizers had risen to 61, most of the flowering plants being grasses; 23 years after the event there were 108 plant species, and 22 years later the score was 276. Some plant arrivals were probably taken there by birds, the seeds either stuck on their feathers or feet, or in their stomachs. Today, much of the islands is covered with jungle.

Enter the human race

When international trade started is not all that clear, but a date around 2000 BC would probably not be far off the mark. Trade in vegetable products almost certainly took place quite early on.

The trade in spices, for example, enabled many colonists to acquire immense wealth up to quite recent times. One of the oldest records of plant collecting that we have is of frankincense (*Boswellia*) trees being brought to Egypt from the land of Punt (Somalia), at the instigation of Queen Hatsepshut about 1495 BC. Dug up with large root balls and packed in wicker baskets, 31 out of 32 survived the journey to be planted at the Queen's great temple of Deir-el-Bahri.

Wherever the Greeks went they took the olive tree with them, while the Romans would not go far without sowing sweet chestnuts. Although the gourmet Lucullus is said to have brought the sweet cherry to Rome from Armenia, where he was busy fighting in 73 and 72 BC, Theophrastus had mentioned it in his great *History of Plants*

The peach was cultivated in China before it was known any further west; it may have come to India, the Near East and, finally, Europe via the Silk Road, or it may have been brought by the Mongols.

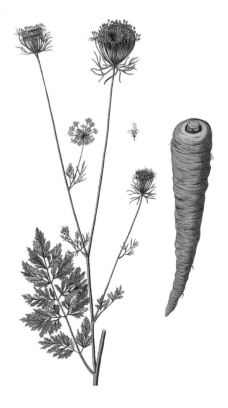

*Wild carrot, **below**, is found in much of Europe but it is only in some areas, such as Afghanistan, that the roots are coloured. It is believed that carrots were first cultivated there, spreading to Asia Minor by the tenth century, Spain by the twelfth, north-west Europe by the fourteenth, and England by the fifteenth. The familiar orange carrot was selected in the Netherlands by about 1600.*

around 300 BC. What seems possible is that the cherry, which originated in Asia, had been domesticated in Asia Minor and Greece long centuries earlier, but that in Armenia larger and sweeter cherries had been selected and cultivated.

Armenia and eastern Turkey lie near the centre of origin of many fruits, and this region is traditionally the home of much fruit development. It was overrun many times, so that its fruits travelled rapidly. The first apples are thought to have been bred there from wild crab apples and the apricot (*Prunus armeniaca*) probably also came to Europe through Armenia.

Apart from invasion, there was deliberate colonization, none more surprising than the great voyages of the Polynesians from AD 750. When they travelled vast distances by canoe – the Hawaiians to Samoa, 2,500 miles, and the Tahitians to the Marquesas, 1,000 miles, for example – they carried roots to plant. Among them, known from tuber remains in archaeological sites, were taro and sweet potato. This is something of a mystery, because the sweet potato is known to originate in Mexico or northern South America. How did the Polynesians have it long before the Spanish invasions? In Peru, sweet potato is called *kuma*: in New Zealand, *kumara*. Is this a proof of Thor Heyerdahl's theory that South Americans travelled to Easter Island – and thence into Polynesia? Or did the Polynesians visit South America and obtain the sweet potato there?

It seems that useful mutations among vegetables have been repeatedly lost. Pliny mentions an African cabbage which sounds like the modern Brussels sprout but which

then disappears from the records. In the fifteenth century there appeared the multi-headed cabbage which, judging from old illustrations, seems to have been half-way between the cabbage and the sprout, but it was not until the late eighteenth century that the Brussels sprout as we know it today arrived.

Pliny's Lacuturnian cabbage sounds like the first hearting cabbage, which did not then reappear until the thirteenth century. The Savoy cabbage seems to have been grown in Germany by 1543, and probably originated in Italy.

People tend to be conservative with new vegetables. Cobbett was recommending American maize or sweet corn in the 1820s, but it was only a century later that it became popular in Europe.

When colonization of the New World began, a number of Old World vegetables and fruits were introduced to the new plantations, while a number of New World vegetables were introduced first to Europe and subsequently to other continents. These included such crops as maize, green beans, tomatoes, cucurbits, peppers and cocoa. An extremely important New World vegetable, the potato took time to become established since the original importation failed to produce sizeable tubers in northern Europe. It was not until at least a century after the original introduction that the potato adapted to cropping in the new conditions and began to assume its pre-eminent position. Surprisingly enough, potatoes did not reach North America from the ancestral home, relatively close, but via the European settlers around 1621.

The apple story

The apple, genus Malus, is certainly the most widely grown fruit in temperate countries, with an annual crop of over 40 million tonnes. The ancestral apple is thought to have originated in south-western Asia, around the conjunction of Russia, Iran and Turkey, where wild apples still grow in quantity.

There are over 50 species altogether. Of these about half a dozen, including Malus baccata, but especially the sweet apple, M. pumila, and the sour M. sylvestris, are among the many ancestors of the modern apple.

People migrating westwards from Asia must have taken apples with them among provisions and thus new colonies of wild trees, often grown from pips, appeared all across Europe. Sizeable fruits were sliced and dried for storing, as indicated by carbonized remains found on many European prehistoric sites.

The earliest records of deliberate cultivation are from the ancient Greeks and Romans, who by inventing the art of grafting – in which growths from a selected tree are "worked" upon wild rootstocks, as **below** – were able to multiply the most desirable of the otherwise one-off seedlings they found growing. The importance of rootstocks in determining the size and vigour of the tree was already appreciated, and dwarf apple trees were grown.

The Romans carried favourites to settlements all over Europe including Britain. After they withdrew, monasteries perpetuated some of their varieties. At least one of these, named "Court Pendu Plat" in the sixteenth century, is still in cultivation.

When Europeans colonized North America they grew their apples almost entirely from seeds. Although this inevitably resulted in a proportion of inferior forms, it meant that American orchards held much greater genetic diversity than those in Europe, where relatively few new seedlings were perpetuated. This has resulted in great differences in the apples grown on each side of the Atlantic.

Malus baccata

Malus sylvestris

Malus pumila

Capsicum annuum

*Coffee, **below**, has been spread round the world, from its centres of origin in Ethiopia and central Africa to Brazil – which it reached via a highly circuitous route through South-east Asia, Amsterdam, Paris and Martinique (see map on page 45).*

Coffea arabica

Capsicum peppers, all of them varieties of one very variable species, were introduced to Europe by the Spaniards and are the New World's main contribution to our spice cabinet.

The rubber tree *(Hevea brasiliensis)* was extracted from a local Brazilian governor by deception. Joseph Hooker at Kew wanted it to send to South-east Asia, but the Brazilians were trying to keep the monopoly; eventually it was on the pretext that the rubber tree seeds were to be a gift to Queen Victoria that permission was given to Henry Wickham to take them out of the country. Earlier importations of plants and seeds had not been successful. Wickham's seeds arrived at Kew in 1876, and germinated there. They were then preserved in Asia, first in the Botanic Garden of Henaratgoda, Sri Lanka, and later at the Singapore Botanic Garden. But it was not until the end of the century that the setting up of plantations began in earnest. The great rubber boom of 1910, which made Malaya the greatest rubber producing country in the world at that time, might never have taken place but for a few dedicated scientists, in particular H. N. Ridley, who was then director of the Singapore Botanic Garden. His book *The Dispersal of Plants throughout the World* still remains the classic work on the subject.

Sir Joseph Banks (1743–1820), then de facto director of Kew Gardens, seems to have been the first to see botanic gardens as a means by which commercial crops could be distributed around the world. The encouragement of new economic crops is still one of the functions of botanic gardens.

It was comparatively easy to transport seeds from place to place: they usually had a high level of successful germination and provided the best method of propagation; even so some types were repeatedly lost.

The real difficulties arose with cultivars. These do not breed true from seed: and with woody species it is necessary to graft a scion of the desired variety onto a rootstock to increase it. The invention of styptics made it possible to preserve scions in good condition, but distance, time and transequatorial change of season between, for example, Europe and Australia, made grafting hazardous. Although some cultivars were introduced from country to country in this way, it proved easier to import seeds and develop new cultivars from them.

Nowadays, with fast ships and aeroplanes this difficulty no longer exists, and one of the more popular modern apple varieties, "Crispin", is a cultivar of Japanese breeding.

Food and medical crops are, not unnaturally, the most travelled members of the plant world, but ornamentals have become increasingly so as human life became less of a battle to survive. From the end of the seventeenth century onwards, the search for ornamentals accelerated, and by the nineteenth century the trade had become so lucrative that nurserymen would pay well for foreign seeds, and those with sufficient capital would employ collectors on expeditions to all parts of the globe.

Sometimes these expeditions had more obvious commercial ends. The Oregon Association sent John Jeffreys in the 1850s to search the Rockies for new trees that might be suitable for forestry work. This followed on from David Douglas's introductions of such plants as Douglas fir, Monterey pine and Sitka Spruce, all of which have become major constituents of forestry throughout the world. Even from Jeffreys, subscribers

received the seeds of many ornamental trees as well as the conifers that it was hoped would prove profitable for forestry.

So far we have talked about the gathering of plant material by collectors as though there were no problems. In fact transport of young plants was extremely difficult. Imagine yourself several months out on a collecting expedition in the Amazon basin; you travel in canoes (there are no roads); it will be months before you get back to civilization. Keeping growing plants alive takes incredible attention. Even the botanist, pressing specimens between sheets of paper, was hard put to it to keep these. On his monumental Amazon journey with the botanist Aimé Bonpland, between 1799 and 1804, the great Alexander von Humboldt wrote that, after travelling six thousand miles, a third of their specimens were found to be destroyed: "We are daily discovering new insects, destructive to paper and plants. Camphor, turpentine, tar,

pitched boards, and other preservatives prove quite unavailing here."

Even on board ship for home, long sea journeys took their toll; apart from anything else plants needed water and it was lack of water for the crew which cost Captain Bligh his 1,000 breadfruit plants *(Artocarpus altilis)* and made him a castaway after the mutiny on the *Bounty*. One might think seed the simplest and lightest means of getting new plants into cultivation, but this was not always so. Seeds might lose their viability and become useless on long journeys (seeds of delphinium and Himalayan primulas, for instance, need to be sown very rapidly after collection, to germinate successfully); while in hot, moist climates they were often attacked by moulds and insects.

The great invention to protect plants was the glass enclosure, known as a Wardian case, invented independently by a Scot, Alan Maconochie, in 1825, and the London Doctor Nathaniel Ward in 1829. Within the

The attempt to bring breadfruit from Tahiti to feed the slaves in the West Indies led to the famous mutiny on the Bounty *(1787): the crew, incensed by thirst and harsh treatment, took over the ship when they saw their water reserved for the plants. In 1793, the vessel* Providence *accomplished the mission.*

Artocarpus altilis

The need for conservation has been recognized only recently. One of the leading botanists of the last century was Sir Joseph Hooker, **above**. *In his* Himalayan Journals, *Hooker recommended the wholesale collection of the spectacular blue orchid* Vanda coerulea, **below**, *as a quick way of making money – without apparently expressing a twinge of concern for the species' survival.*

sealed cases plants could survive for years if need be, safe from drying heat, wind and salt spray, in a climate of their own creation, and needing very little extra water since transpired moisture returned to the soil. On a larger scale, greenhouses are now essential to both horticulture and agriculture.

Like virtually everyone else, the plant collectors had no inkling of any need for conservation. Two of the worst offenders seem to have been Sir Joseph Hooker, in the Himalayas, and Benedict Roezl, who brought back 100,000 orchids from one region of Mexico alone. In comparatively recent times the Vesubie valley in southern France was denuded of *Lilium pomponium*, and northern Portugal of most of its wild narcissi.

Wandering weeds

Plants may always have travelled, but unfortunately, those that travel most today, with the aid of people, are what we call weeds. Although many seeds are adapted for transport by animals, once humans started moving the opportunities were even greater. Seeds travelled on clothing, blankets, covering, among imported seeds and roots, in mud on shoes, in the digestive tract; they moved on wheels and vehicles of every kind. Many plants, deliberately grown as crops or ornamentals, escaped into the surroundings with devastating results.

Sometimes humble garden nuisances can colonize very readily. The dandelion, imported accidentally in grass seed from Europe to southern South America, now grows like a native on the pampas and is a favourite food of the rhea. The plantain,

brought by settlers from Europe into North America, spread so rapidly that the Native Americans called it Englishman's foot.

Introduced plants, growing in isolation from their own finely tuned ecological habitats, often get out of hand because there are no predators available to control them. An example of this is found on Mauritius, the one-time home of the dodo – the very symbol of extinction. Much of the indigenous vegetation has been degraded by vigorous introduced species such as strawberry guava (*Psidium cattleianum*), encouraged by disturbance of the habitat due to alien animals such as pigs, rats and monkeys.

One of the saddest examples of invasion by exotics is in the Cape of South Africa, comprising an area made up of fragments totalling less than the area of the Kruger National Park. The Cape is so rich that it has been called the Cape Floristic Kingdom; of its 9,600 species of plants, about 70 percent are endemic. There are only six such kingdoms in the world; the biggest, the Boreal Kingdom, covers most of the northern hemisphere north of the great deserts. Much of the Cape has been ploughed up for growing crops, and countless hectares have been choked by invasive plants introduced by settlers from other countries.

These tales of opportunism could be multiplied many times. It sometimes seems as though these wandering weeds, taking over not just areas where habitats have been destroyed by humans but also, in some cases, entire mature plant communities where they find the conditions specially favourable, could one day be the inheritors of the earth's surface.

Prickly pears, **left**, brought into Australia late last century as garden curiosities, found the conditions so perfect that by 1925 they occupied over 24 million hectares. They were eventually eradicated by a cactus-eating moth from Argentina. In the last few decades

lantana, **above**, has taken over the habitats of many endemic species in Hawaii. The beautiful South American water hyacinth, **top**, has clogged tropical waterways worldwide, and invaded paddyfields in South-east Asia.

CHAPTER FOUR

Our Daily Bread

The range of plants known to have been used by humans as food, drink or flavourings is astonishing – as many as 10,000 species including fungi and seaweeds have been recorded and we assume early peoples must have tried many more. More than 50,000 species have been recorded as edible in some shape or form, and at least 80 percent of the world's food is derived from plants. It is hard for us today to conceive the human cost of identifying this wealth of natural resources. The great majority were probably first consumed by hunter-gatherer tribes, living off the land in isolated, local communities. Often these people must have died or suffered as they learned by trial and error not only which plants but also which part of them – the fruit, flower, leaves, stem or root – were safe, poisonous, or edible only after culinary treatment.

Many of these plants are still eaten from the wild but only about 3,000 seem ever to have achieved wide acceptance, and only about 150 have been cultivated to any extent. During the long centuries of selection for the heaviest seed-head, the most useful food, the sturdiest plants, our modern crops have gradually evolved, and certain favoured staples have come to dominate the global menu. Today we rely on about 20 species to provide most of the world's food needs, and over half of our calorific intake is supplied by just three grasses: wheat, rice and maize.

Over the last fifty years, the process of specialization has accelerated, fuelled by genetic research, until an increasingly large share of our harvest rests on certain carefully bred, high-yield strains designed to feed a

Rice – here being planted in a paddy field in Bali, Indonesia – is one of the "big three" cereals that have become world staples.

hungry world. The more highly bred and artificially cultivated a crop, the more prone it is to attack by new strains of disease or pest; a third of all food produced is destroyed by pests and diseases.

To keep one jump ahead in this leapfrog struggle, new resistant strains are constantly needed. The key to maintaining this improvement lies in the stock of three types of plant, all vital for the future. First are the very diverse traditional cultivars – landraces – many of which are being replaced by the modern, uniform new crops produced by crop breeding and also by biotechnology. Modern methods of genetic modification (GM) have come in for much publicity and debate, some of it very emotional, and not all of it well-informed. As with many scientific advances, there are both benefits and problems with GM, as discussed elsewhere (see Chapter 10, page 171). Second are those wild relatives of crops which are still to be found in their undisturbed natural habitat; wild crop relatives have played an increasingly significant and fascinating role in crop development, especially in breeding disease resistance into the modern strains. The third potential resource for the future, also threatened, is the unused potential of the other 10,000-plus species of which we have inherited knowledge. Never has the need for exploring new possibilities been greater; these plants may be valuable in cultivation, cross-breeding and genetic improvement, and also have potential as new staples or useful foods.

Virtually all the staple foods we know today have been cultivated since antiquity. It is surprising we have added so little to our

World arable land use

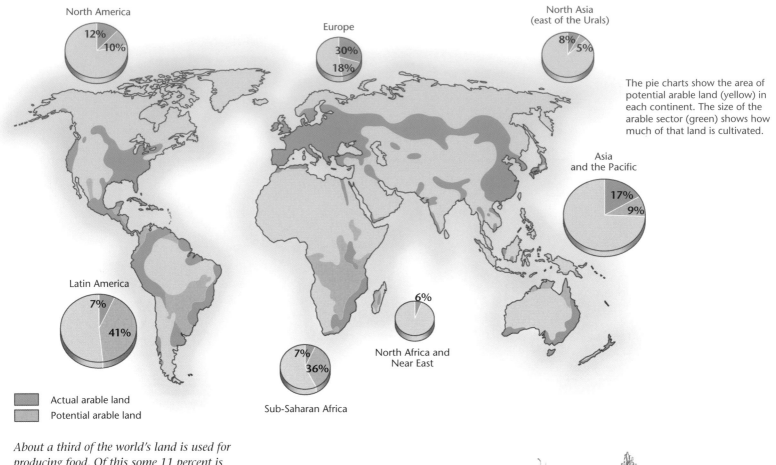

North America
12%
10%

Europe
30%
18%

North Asia
(east of the Urals)
8%
5%

The pie charts show the area of potential arable land (yellow) in each continent. The size of the arable sector (green) shows how much of that land is cultivated.

Asia
and the Pacific
17%
9%

Latin America
7%
41%

North Africa and
Near East
6%

Sub-Saharan Africa
7%
36%

Actual arable land
Potential arable land

*About a third of the world's land is used for producing food. Of this some 11 percent is arable, 0.6 percent carries various permanent crops from orchards to plantations, and 23 percent is permanent pastureland, mostly for meat-yielding animals. About 30 percent is designated as forest and woodland – and some of this is being rapidly destroyed. As the map, **above**, shows, the southern continents, though large in area, have a lower percentage of good arable land for crops than some northern ones. Drought and poor soils also affect many southern croplands, and the effect on their production is evident from the second map, **facing page**. The world's cereal harvest is heavily dominated by the north. Moreover, increased yields – a gift of the new crop varieties, fertilizers, and technology of the Green Revolution (p. 66) – have barely touched Africa, and only moderately helped South America. In Africa, particularly, crop improvement is lagging.*

Maize   Sorghum   Millet   Barley

Cereal grains are the major source of concentrated carbohydrate for both food and fodder. After the "big three" – wheat, rice and maize – the most important are barley, sorghum, oats, millet and rye. Which cereal is dominant where is determined by climate, and influenced by local preference. Barley is tolerant of most conditions and can grow from the Arctic to the Andes, from the Himalayas to Ethiopia, including soils too saline for other cereals. Sorghum and millet are both staples in the drier regions of Africa and Asia. Sorghum is particularly hardy and drought-resistant: it is the predominant cereal in areas too hot and dry for maize. China and India are the major millet-producing countries.

Other cereals of local importance include teff (Eragrostis abyssinica) of Ethiopia, and Job's tears or adlay (Coix lachryma-jobi) of India which has much as yet unrealized food potential (p. 75).

A few plants not of the grass family produce cereal-like grains: amaranth (Amaranthus *spp.*), *widely grown in the Andes and South-east Asia; quinoa* (Chenopodium quinoa), *a declining crop in the Americas; and buckwheat* (Fagopyrum esculentum), *widely grown in Canada and Russia.*

In virtually all undeveloped countries, winnowing chaff from grain is carried out by hand.

World cereal production
(million tonnes per year)

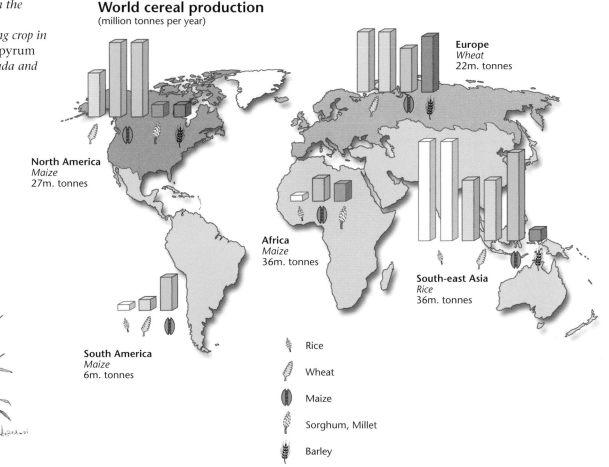

Europe
Wheat
22m. tonnes

North America
Maize
27m. tonnes

Africa
Maize
36m. tonnes

South-east Asia
Rice
36m. tonnes

South America
Maize
6m. tonnes

- Rice
- Wheat
- Maize
- Sorghum, Millet
- Barley

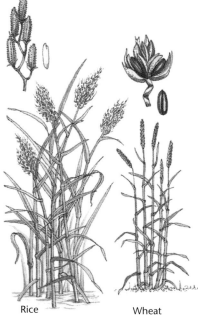

Rice   Wheat

55

Dioscorea alata

Yanomami Indians of the Amazon harvesting cassava for a feast.

Apart from the almost ubiquitous potato, world production of starchy roots and tubers divides approximately into temperate and tropical. Brazil is the major producer of cassava or manioc (Manihot esculenta), seen being gathered by Amazon Indians **above right**, while China dominates the sweet potato (Ipomoea batatas) *league*. The greater yam (Dioscorea alata), **above left**, originated in South-east Asia and became the most widely distributed of all yams. Their importance, however, is gradually being eroded by the less nutritious cassava or by cocoyams in some areas.

Potato (Solanum tuberosum) *is the most important non-cereal world food plant; production accounts for over half the annual tonnage of all starchy roots and tubers. Russia and Poland are world leaders in potato growing. It now grows worldwide except in the lowland humid tropics; but the very tasty, small tubers of the potato's Andean homeland are rapidly declining in availability.*

Other temperate roots are comparatively unimportant, though sugar beet, a fairly recent crop, now accounts for almost half the world's production of sugar.

World root, tuber and pulses production
(million tonnes per year)

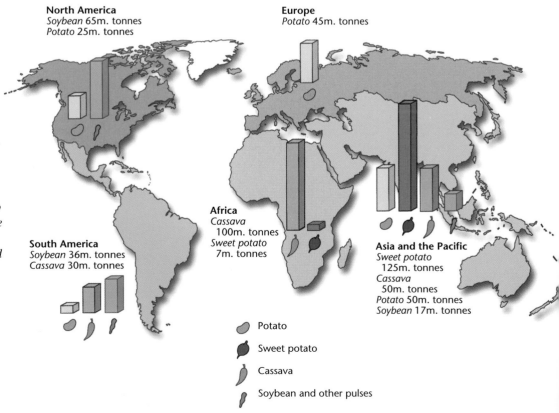

North America
Soybean 65m. tonnes
Potato 25m. tonnes

Europe
Potato 45m. tonnes

Africa
Cassava
100m. tonnes
Sweet potato
7m. tonnes

South America
Soybean 36m. tonnes
Cassava 30m. tonnes

Asia and the Pacific
Sweet potato
125m. tonnes
Cassava
50m. tonnes
Potato 50m. tonnes
Soybean 17m. tonnes

Potato

Sweet potato

Cassava

Soybean and other pulses

Lens esculenta

The soybean (Glycine max) *is not the most appetising of the pulses but it outstrips the others both in quantity produced and variety of uses including new "meat analogues". Its vast expansion as a western crop only occurred 50-odd years ago. It overtook wheat and maize as the most important US cash crop in 1973. However, its future potential is threatened, since many of its wild relatives are being destroyed. Brazil is now the main exporter of soybeans and soybean products, followed by Argentina and the USA, and China is also a major producer.*

India is the major producer of lentils (Lens esculenta), **above** *– one of the oldest and most popular of Old World crops. Tasty and nutritious, they contain 25 percent protein. Other important producers are the USA, Australia, Canada, Pakistan, Syria, Argentina, Chile, Turkey, Ethiopia and Spain.*

A minor leguminous crop is the carob (Ceratonia siliqua), *a tree producing large pods rich in sugar and protein, eaten raw, made into alcoholic beverages and a "health-food" substitute for chocolate, or used as fodder: its seeds were the original carats or jewellers' weights.*

ancestors' choices: a great innate conservatism seems to have been at work, since useful food plants will perform admirably away from their natural homes under cultivation when given the chance. Once people get used to a few kinds of food as major parts of their diet, they tend to stop experimenting.

The staple foods come from a small range of families: there are some 460 families, yet the 130 most important food plants fall into only 32 families. The most important are the grass family, Poaceae (Gramineae) – rice, wheat, maize, sorghum, barley and sugar cane; Solanaceae – potato; Convolvulaceae – sweet potato; Euphorbiaceae – cassava or manioc; Leguminosae – beans, pulses, soybeans and peanuts; Palmae – coconut; Musaceae – bananas; and Chenopodiaceae – sugar beet.

Wheat is the most important crop in the world: the principal food of 35 percent of the world's population, nearly 600 million tonnes are produced annually on more than 200 million hectares of land, providing about 20 percent of the world's calorie consumption as well as a large proportion of all our essential nutrients. Rice is the leading tropical crop – even more important than wheat in terms of direct consumption – while maize ("corn" in North America) is staple in many tropical and subtropical regions. Although China is traditionally regarded as a rice-eating nation, in fact wheat flour is in high demand, mainly for making noodles, and China is now the world's biggest producer of wheat.

Fleshy roots, tubers and rhizomes, collectively known as root crops, are one of

nature's ways of ensuring continuity for perennial plants over dry or cold seasons all over the world. This made them naturals for early people to exploit, particularly in tropical lands. They are vitally important providers of food energy for populations throughout the world – especially the potato and the tropical yams, cassavas, cocoyams and sweet potato. Carrots, parsnips, swedes and turnips also feature prominently.

Pulses are the hard, dried, edible seeds of the pea family, Fabaceae (Leguminosae), and a most important source of protein; in the tropics grain legumes are often the only source available to supplement the carbohydrate crops, while forage legumes such as clover and lucerne provide high protein diet for livestock. These useful plants have root nodules containing nitrogen-fixing bacteria and are therefore important in maintaining soil fertility, especially in tropical areas of nitrogen-deficient soil. Often deep-rooted, legumes are further valuable in aerating the soil and encouraging rain to permeate.

Many plants produce oil reserves rather than starch, but only about 30 species have been used for oil extraction, and of these a dozen or so account for 90 percent of the vegetable oil supply. The olive tree is characteristic of the Mediterranean region, where there are probably 700 million trees, covering about 9,500,000 hectares. Olives may live for 1,000 years, with ages of 2,000 years claimed for some specimens. Numerous references in the Bible confirm that the olive was one of the earliest trees to be grafted.

The African oil palm, very widely cultivated in the tropics, gives a higher oil yield per hectare than any other oil plant. It

Citrus aurantium

Peanuts (Arachis hypogaea) *are grown throughout the warmer parts of the world and, except for about 5 percent, all are consumed or processed locally. The high (30 percent) protein content of the Virginia type makes it a valuable source of energy in the diet, while the very high (47–50 percent) oil content of the Spanish-Valencia type is more suitable for oil production. The flowers grow above ground in the orthodox way, but after pollination the developing seeds – the ground-nuts – are pushed underground. This old illustration,* **above***, pre-dates the discovery of this fascinating self-sowing mechanism.*

The olive tree (Olea europaea) *seems to have originated in the eastern Mediterranean but as early as the fourth millennium* BC *variable hybrids certainly existed, as kernels found at Byblos confirm, and the olive ancestor is not known. From its eastern origin it has spread westwards to Greece, Italy, North Africa, and to Spain. Currently some one and a half million tonnes of oil are produced annually. Olive oil has been one of the great foods of this historic area but production is declining now since harvesting depends so much on manual labour: one can still see the olives being beaten down with poles onto sheets on the ground,* **far right***, just as ancient Greek vases depict,* **right***, or fallen ones picked off the soil by hand.*

Another important fruit group, Citrus, has many manifestations – from oranges, sweet **(above)** *and sour, mandarins, tangerines and their like, to grapefruit, lemons and limes, and the less familiar pummelo (p. 75). Their wild relatives are endangered by the destruction of Asian forests.*

The modern cabbage (a cultivar of Brassica oleracea)*,* **above,** *with its mass of tight-packed leaves, developed from wild ancestors with only a few tough leaves. Such tight-headed varieties were already being cultivated in the twelfth century.*

produces two quite different types of oil, that from the palm kernels, used for making margarine and soap, and that from the fleshy part of the fruit used more widely for industrial processes. This is one of the most rapidly expanding plantation crops today.

Sunflower is the world's second largest vegetable oil crop. Some seeds are eaten, but most often oil is made from them, and the remaining seed cake is valuable as stockfeed. Enormous fields of bright sunflowers are perhaps the most attractive crops of any.

Sesame has been claimed as the oil seed most anciently used; there are numerous records of its use in cooking, medicine and cosmetics, and for ritual. It is a fast-maturing annual and its seeds can contain 60 percent very nutritious oil. Yields vary but more attention to breeding sesame could greatly improve its standing among the world's oil supplies.

Coconuts are multi-purpose palms of humid tropics, capable of producing food from their fresh nut meat and "milk"; copra from the dried meat, used for stockfeed; oil pressed from the copra, mainly used for soap; and coir fibre from the husks.

Nuts, other than the leguminous peanut, are minor crops, but nonetheless quite widely grown in some areas. They are borne on trees or shrubs: sweet chestnut, walnut, hazel and filbert, pecan, cashew, pistachio, and macadamia nut – the last being one of the few useful native Australian food plants, now becoming widely grown and the subject of recent breeding.

Fruit must always have been among the items which people have added to their basic diet to lend taste and interest and, in partic-ular, to satisfy a desire for sweetness. Some fruits became important very early on. Two of these are the grape – largely because of its capacity to be made into wine – and the date, which was the staple of North Africa and the Middle East. Dates are now also cultivated in East and South Africa, in south-west USA and in Central and South America, and even in southern Europe, with a world population of about 105 million trees. More is said about these fruits in Chapter 9.

The most widely grown tropical fruit is the banana, which arose in South-east Asia. Modern varieties are almost entirely seedless so new varieties are bred from semi-wild types and arise also by the numerous muta-tions which constantly appear. Pineapple is another important tropical crop, mainly grown for canning. It is unusual among fruits in not growing on a tree or bush but from a spiny rosette – it is a bromeliad like the indoor urn plant – the fruit being an oval thickening of the floral axis, with the seeds embedded in the outside skin.

A fruit beloved of animals is the South and Central American avocado, intriguing among tree fruits in being savoury rather than sweet. It contains much protein and is even eaten by carnivores like jaguars. It was probably eaten by people as early as 7000 BC, and cultivated by 5000 BC. Widely grown, there are many distinct varieties, though these are seldom named to the consumer.

Tomato is usually used as a vegetable and referred to as such, though the part we eat is the fruit. In western countries it has become a very important crop and in the US its annual value is barely below that of potatoes. Tomatoes have a flavour different from that of any other vegetable and are rich in vitamins, notably vitamin C. The tomato tagged along as a weed in maize and bean fields for centuries. Like the potato, the tomato was introduced to Europe by the Spanish invaders of South and Central America, though only a tiny proportion of the potential varieties were brought across the Atlantic, and only in this century have other species that have become important for pest and disease resistance been used for breeding.

Green leaves and sometimes shoots or stems are very important additions to diet, mostly in providing protein and a little fibre, but especially calcium and vitamins often lacking in other foods. Sadly the intake of leaves is far less in many poor tropical coun-tries, where it is almost essential for well-being, than in richer temperate ones even though supplies, often plentiful, can be obtained from the "spare" parts of root crops like cassava. However, in some regions, notably in Thailand and southern China, many leaves and even flowers feature large in the local traditional diet.

The cabbage tribe, grown in temperate regions, deserves special mention because of its extraordinary variety. Cabbage began as a woody plant with a few loose leaves. In its current forms numerous tightly packed leaves surrounding a terminal bud give us cabbages, enlarged side buds of packed leaves on a long stem Brussels sprouts, enlarged flower buds cauliflowers and various broccoli, and thickened stems kales and the turnip-like above-ground kohl rabi.

Onion and its relatives have been recognized since very ancient times as providing a unique flavour; some like garlic and chives are more important as flavouring than as basic food. Their medicinal and tonic value was also appreciated early on. There are references to onions, garlic and leeks as early as the first Egyptian dynasty (3200 BC) and many in the Bible, while garlic features in early Chinese records. Besides the globular bulb onion and smaller shallot, and the elongated leek and green-leaved Egyptian kurrat, eaten as vegetables, there are several kinds of bunching onion which are eaten green – in northern China one can see people gathering wild forms of these to liven up their meals.

Differences in diet

More than enough grain is grown annually to feed the whole world. The problem is not only one of its distribution but of the way in which we use it. Dietary habits lie at the root of many of our troubles. In the west a great deal of meat is consumed, and meat-producing animals need feeding. In the United States about 70 percent of the grain harvested is now fed to these animals, apart from vast areas of grass and other fodder crops. A westerner eats about 65 kg of grain a year, plus meat enough to account for a further 900 kg of grain. By comparison, the average Chinese eats 160 kg of grain products and under 20 kg of meat. Foods considered necessary to feed 200 million in the west would be adequate for 1,500 million Chinese.

Big business creates associated problems. In Europe production from dairy farming for example exceeds demand so that the Euro-pean Union currently staggers under a "butter mountain" and is awash with a "milk lake" engendered by a price support system. The pastoral idyll of cattle grazing in a flower-filled meadow is often far from reality. The stock are pastured on near monocultures of ryegrass inimical to wildlife, or reared on feeds of cereal, soya, manioc, or even surplus milk products. Cattle in particular are often unhealthy and short-lived under these modern feeding regimes.

In the Far East where milk has not traditionally been part of the diet adults do not retain the ability to digest it, so dairy farming is hardly practised. For the majority of people in the tropics the amount of meat, eggs, milk and fish, as well as fat and sugar in the diet, is low. The staff of life tends to be cheap, stomach-filling plant food, like boiled rice, cassava or sweet potato, often rather boring to more sophisticated palates. Often few green vegetables are eaten, although this varies with the region.

Deficiency in vitamin A, which is provided ready-made in milk, eggs and liver, but which otherwise the body manufactures from the pigment carotene found in abundance in green leaves, causes among other things degeneration of the eye surfaces that can result in destruction of the cornea. Between 100 and 140 million children are vitamin A deficient, mainly in Africa and South-east Asia, and as many as 500,000 become blind every year as a result. It is dreadful that they should so suffer when one leaf of taro is sufficient to cover the daily needs of three toddlers, and two teaspoons of cooked fresh greens a day could protect a child from a life of blindness.

Many greens could be gathered free to supply this requirement. In some European countries there is a widespread habit of collecting wild greens, often the tops of assorted meadow or roadside plants. Chickweed, white dead-nettle and stinging nettle are all prized by the initiated. Basic tropical crops, such as cassava and sweet potato, grown for their tuberous roots, pawpaw for its fruit and milky latex, and jute for its fibre, can supply the all-important green component, and if leaves are picked in moderation this will not affect the main crop. The protein component of leaves is important where food intake is marginal, and fresh greens supply many other important components of diet, such as vitamin C, iron, folic acid and calcium. In China, this problem is recognized: traditionally every family grows its own plot of greens to add to their rice. This is just as well, for in their largely vegetarian diet well over 90 percent of calories are provided by cereals, and most proteins from cereals and legumes; a great deal of the cereal is rice. In its unrefined state this is a good food source comparing well with wheat, but unmilled brown rice is often scarce or expensive. In China, as elsewhere in Asia, white rice, milled and polished, is preferred. All the nutritious coloured pericarp is removed as bran; the protein content decreases from ten to seven per cent in polishing, and washing and cooking methods deplete the nutrients further. A diet too heavily dependent on refined rice can result in deficiency diseases such as beri beri.

There is growing evidence, too, that western foods can be damaging to health.

Contrasts in cultures in the marketplace
Leaves strewn on the roadway are the main items on sale on this Calcutta riverside, **above**, *seen at sunrise.*

A greengrocer's stall in Iraklion, Crete, **left,** *offers a wide range of fruit and vegetables, nicely presented – almost all grown in the comfortable climate of this mid-Mediterranean island.*

Despite the apparent barrenness of their homeland, Bushmen of the Kalahari Desert find enough sustenance here. They are among the few surviving hunter-gatherers, following a lifestyle which began for emergent Homo sapiens *around 20,000 years ago. Besides plants they eat insects and whatever other animals share their harsh environment.*

Some scientists say that the food we eat, especially overconsumption of so-called "junk food", is a major cause of death and disease, and that we should dramatically reduce our intake of sugar, salt, and animal and dairy fats – the modern killers. Many rural Africans, like the Dogon tribe, are not prey to such problems as coronary heart disease or cancer of the bowel for instance, and one of the main differences between the diets is the amount of fibre in the food. Dietary fibre is the indigestible part of plants such as bran, which western food refiners remove so that, for example, bread is spongy, light and white, not a healthy, heavy, brown chew. Fibre aids digestion and the rapid passage of matter through the body; it thus prevents cancer of the bowel by quickly flushing poisonous agents out of the gut. The low proportion of roughage that many of the elderly of Britain consume means that food may take *two weeks* to pass through the alimentary tract.

Properly used, plants can supply all our carbohydrates, protein, edible fats and oils, vitamins and minerals, and feed our animals. We need all these components of food and a mixed diet is absolutely essential.

The beginnings of cultivation

Primitive people operated as hunter-gatherers from perhaps 20,000 BC to 8000 BC, killing wild animals for food, fishing, collecting molluscs, and gathering all manner of edible plant material. This way of life still exists in certain communities: Australian Aborigines, Kalahari Bushmen and some central African tribes do not attempt any cultivation.

Gathering nature's gifts seems a carefree life, and one might wonder why ancient peoples turned to agriculture. In fact, hunter-gathering is not always very comfortable. The Aborigines, for instance, were always faced with a harsh environment with scanty resources.

As we saw in Chapter 3, hunter-gathering means keeping on the move. Settling down in one area was possible if small numbers of people inhabited rich food areas, especially if there was good fishing adjacent from sea or lake. And once one has settled, there is first of all the possibility of stockpiling food over cold winters or dry seasons.

The earliest record of settlement combined with such storage is of the Natufians of Palestine, who dug store pits which they lined with plaster and had grinding stones and sickle blades. Around 9000 BC they began to gather cereal seeds to augment their diet of wild gazelle and other game. Such a settled lifestyle is a prerequisite for agriculture to begin, the settlers becoming familiar with their plants at all seasons and, in due time, able to tend growing crops consistently.

There is little doubt that crop cultivation developed independently in several places. For a long time the "cradle of agriculture" was thought to be sited in the "Fertile Crescent" of the Near East; this was largely because there were in that area accessible, well-preserved archaeological remains of stone or mud constructions, many of Biblical significance, which were among the earliest to be investigated.

Barley and wheat were both cultivated in ancient Egypt from the fifth millennium BC, wheat becoming the more important of the two cereals. These tomb paintings from the Eighteenth Dynasty (1567–1320 BC) show how the ripening wheat crop was sized up by inspectors with measuring lines, who set the government's tax quota. After harvesting, the crop went to the granary for final tallying.

In these dry conditions evidence of early agriculture has come from seed imprints on pottery, grains preserved in ancient caves and tombs, or carbonized near fires. Such grains have been radiocarbon-dated as 9,000 years old. Archaeological evidence equally confirms that northern China was the centre for an agricultural system 7,000 years ago: it was based on foxtail millet, which by 2700 BC came to be held sacred.

Squashes, beans and maize were cultivated – in an ingenious self-balancing triple-crop system in the same patch of ground – in Central America at the same time; while a system based around potatoes, cultivated with digging sticks, started up independently in the South American Andes some 8,000 years back. In warmer areas of South America, cassava was clearly being cultivated as early as the third millennium BC since there is good evidence of trade in cassava flour.

Despite the liability of succulent fruits and roots to decay in hot moist areas, radiocarbon-dated remains in South-east Asia suggest that yams were part of a vegecultural civilization (Hoabinhian) in 8000 BC. Yams are a vital part of early agricultural history since different species exist in Asia, Africa and tropical America, and formed the basis of early vegeculture in these areas (in America rather overshadowed by cassava) and are still vital to subsistence agriculture in many areas. Yams are cultivated from small pieces of tubers or by stem cuttings, and their propensity for rapid rooting must have been noticed early on as people gathered them and left odd bits in the ground. Curiously enough cultivated

yams seldom set seed, so that many varieties have been increased for thousands of years by vegetative propagation.

Hypotheses of the domestication of cereals are very different. Best known is the rubbish-heap hypothesis of how cultivation began. It describes a scenario of rubbish piled up beside the homesteads – fruit cores, seeds, bones, offal and excreta – all decayed into rich compost. Some of the seeds would germinate in the fertile ground, and the plants that the gleaners had trudged many weary miles to collect would suddenly be there on the doorstep, a gift from the gods. The gatherers would have collected proportionately more of the bigger and least fragile wild produce, so the plants on the heap would already be diverging genetically from the wild. The people gradually discovered that some examples of plants produced better seeds or fruits than others so were worth deliberately storing for later sowing and harvesting. And weed plants that could not survive in the fierce competition of the

A coin of Cunobelin, king of the Belgic Catuvellauni, who invaded Britain in 75 BC, made his capital at Colchester and had an alliance with Rome. It shows an ear of cereal, probably barley which was the most abundant grain in cultivation in Britain from Neolithic to Viking times.

Greengage, Coe's golden drop

Plums and their relatives have a complex ancestry and probably originated in central Asia. The delectable greengage, Coe's golden drop, **above**, *was raised in 1800. Like most plums today it probably derived from the damson plum* (Prunus institia), *the cherry plum* (P. cerasifera), *and the sloe or blackthorn* (P. spinosa), **below**. *The spininess of many ancestral forms is now lost in cultivated kinds. Cultivated plums have six sets of chromosomes; the wild plums two.*

Blackthorn *(Prunus spinosa)*

primeval woodlands beyond the settlement would luxuriate on the kitchen middens. Some, such as amaranths, legumes, squashes and cucumbers, had potential as food crops; others like nightshade turned out to be deadly poisons – but even poisons have their uses.

How crops develop

We can begin to see a pattern emerging from historical obscurity. All kinds of wild plants are tried out for food; some prove better than others. Then some ingratiate themselves with people and thus become prime candidates for cultivation, though they are not deliberately chosen.

If one places the wild species and the crop derivative together for comparison, one of the most noticeable things is that the crop plant is much bigger, and not always just in the character desired, such as the grain in wheat, root in carrot, or fruit in apple, but in other parts as well. Thus the sunflower is bigger flowered, bigger leaved and thicker stemmed than its wild relations. Crop plants may change proportions: the peanut that splays its nut-bearing stems metres away from it, in the wild neatly burying them in the ground after flowering, gathers them round its centre when domesticated. The garden cabbage has a firm solid heart – its wild ancestor a few loose leaves. Wild wheat ears are brittle, shattering at a touch when mature to give the seeds a chance of being spread around. The pieces have long "beards", hard to remove from the seeds, which aid them to winkle their way into cracks in the soil,

ready to germinate when rain comes. Early gatherers would have favoured any wheat which had mutated so as not to shatter when harvested.

How have these changes occurred? Early cultivators gradually realized that there were bigger grains of wheat than normal, and by and by non-brittle ears; larger yam tubers, more prolific peanuts, sweeter apples. The "better" forms were sown or replanted so that gradually there was an improvement in the crop by the process of selection.

During such early stages of selection natural breeding between closely related species could occur, say, in the wheatfield, resulting effectively in quite new plants. Often there also occurred unpredictable internal changes in plant cells, known as mutations or sports, which give rise to new forms. One common change is the doubling of the number of chromosomes from the standard double set (diploid) to a quadruple set (tetraploid), which very often creates a larger and more vigorous plant. This was for instance very important in the development of wheat.

A cultivator with a sharp eye would pick out such novelties and grow them on separately. Until the very recent achievement of deliberate cross-breeding the only way that people improved their plants was selection of the best in a crop, with special reference to unusual, fortuitous seedlings or sports. In the same way unwanted characters, for instance natural plant defences like spininess in the wild plum and pear, or bitterness and toxicity in the wild potato tuber, were "selected out" early in the course of their gradual domestication.

From the wild sunflower (Helianthus annuus) **above**, *with its flimsy stems and relatively small flowers and few, small seeds, have been selected the massive forms now grown commercially for their oil-rich seeds, as in Provence, France,* **top**.

Clearly, crops develop a wide range of physiological adaptations as they are grown in unnaturally cultivated and fertilized soil, and later in new habitats as people transport them. Many become able to survive in conditions quite different from those of their original homes. The progenitors of maize are plants which time their lives according to the sun's short-day regime in tropical climes; it has since been transformed into a plant that will grow in the long-day cycles of temperate climates. The same is true of the potato from equatorial South America.

The process of adaptation means that many crop plants may become quite unfit to compete in the wild with their original congeners: they may lose their power of natural dispersal as has occurred in non-shattering wheat; the shortening of the peanut flower stems to convenient length for cultivation equally reduces its ability to spread successfully. Selection for rapid, even germination of seeds puts all the eggs into one basket: the wild plants hedge their bets by staggering germination.

Plant breeding

Modern agriculture hinges on purposeful plant breeding. This has been carried out only during the last 200 years and was not possible on any scientific basis until the monk Gregor Mendel worked out the principles of inheritance in 1865 (work which was neglected until the early twentieth century). Certainly the more advanced methods of maintaining or improving soil fertility, and mechanization, have had their effects; but without breeding, the high-yielding crops responsive to high fertility would not have been produced, nor would the thick-skinned cultivars that can put up with mechanical harvesting and transport to markets (even if this often makes them less palatable, as with tomatoes).

Plant breeders aim to improve yield and quality, and have various methods of achieving this. Given a crop with a fair amount of natural variation, a great deal can be done with a keen eye, a rabbit's tail for pollination and a prayer; but modern

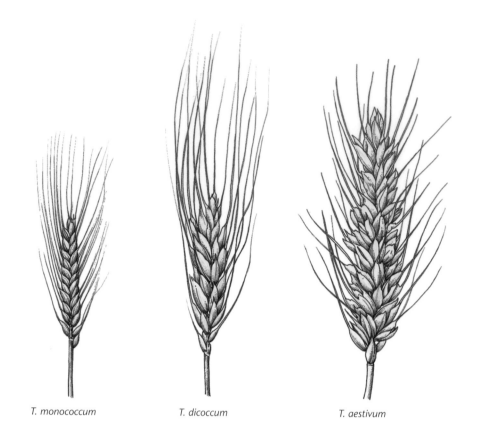

T. monococcum   T. dicoccum   T. aestivum

Ancient cereal grasses had heads which shattered readily so that the seeds would scatter widely, and long "beards" which, caught by the wind, helped the seeds force their way into the soil. Accidental hybridization and, in particular, mutations produced some varieties with non-shattering heads and more easily removed beards; it was these which primitive people selected, first for harvesting from the wild and later for deliberate cultivation. Once their natural mechanism for seed dispersal is lost, the plants become dependent on their cultivators for survival. Among the earliest domesticated wheats were einkorn (Triticum monococcum), **far left,** *a diploid (with two sets of chromosomes) once widely grown in Turkey and Europe; and emmer (T. dicoccum),* **centre,** *a natural tetraploid (four sets of chromosomes) once grown in the Near East, Africa and Europe, together with another wheat,* T. timopheevii. *Today's bread wheat (T. aestivum),* **left,** *is a hexaploid (six sets of chromosomes).*

breeders have a new range of tactics – irradiation, chromosome doubling, tissue culture, and so on – to encourage "unwilling" plants to breed and then to sort the progeny.

One of the landmarks of plant breeding this century has been the development of triticale – the first time a crop has been synthesized by hybridizing species from different genera, wheat (*Triticum aestivum* or *T. turgidum*) and rye (*Secale cereale*). Natural hybrids of the two have frequently been reported, but are usually sterile, and so are non-starters as cereal crops. It was only with the improvement of embryo culture and chromosome doubling techniques in the 1930s that work began in earnest, aiming to combine the winter-hardiness of rye with the high-yielding properties of wheat. The result resembles wheat in appearance, but it is hardier, and in poor, parched soils will out-yield wheat. It is also better nutritionally, with higher content of protein and essential amino acids. More than half a million hectares are planted each year with triticale.

At present it is used mostly as animal feed and experimentally certain lines of it have produced higher growth rates in rats than other cereal grains. It also has potential for alcohol production as it has a high starch content. But hopes are high that it will prove its worth as human food, not only because of its nutritional quality, but because it grows better than wheat on marginal land.

Rice is a spectacular example of breeding for high yields. It was being hybridized as early as 1906. Breeding in Japan and the United States and, finally, in combination, steadily improved the cropping capacity of rice until, in the late 1960s, several "semi-dwarfs" appeared – compact, stiff and erect, with plenty of sideshoots, early maturity, no sensitivity to day length, and responsive to high amounts of nitrogen fertilizer. Breeding initially achieved considerable resistance to major pests and diseases, but now these are outpacing the breeders' efforts. The new rice varieties replaced long-strawed ones, tradi-tional in Asia, adapted to grow above the weeds and floods (breeding to increase the

weight of grain on such plants would have caused them to keel over).

The use of these new high-yielding rice and also wheat varieties in Asia has given rise to what is called the "Green Revolu-tion". This, alas, has shown that plant breeding for higher yields is not always a success, at any rate in human terms. The amount of wheat and rice produced in developing countries has certainly been greatly increased but at high social cost, and also loss of genetic resources.

The revolution has become like a pyramid of cards with the high-yielding crops balanced precariously on top. Remove the fertilizer, or the irrigation water, and the stack collapses. Further inputs are needed, and also herbicides to suppress the weeds, fungicides to quell disease epidemics that attack the vast acreages, and machinery to apply the pest-killers and fertilizers as well as to plant and harvest.

The high-yielding wheats were produced by the International Wheat and Maize Improvement Centre in Mexico. This was founded on money from the Rockefeller Foundation which teamed up with Ford to found the International Rice Research Institute in the Philippines which produced the miracle rice. Multinational corporations were in on the start of the Green Revolution and, by supplying the chemicals, seeds and machinery, have continued to profit from it since. An example of the social consequences of such activity is Bangladesh where the introduction of combine harvesters has taken work and income from a great many pairs of ill-fed hands.

Often the rural local farmer does not benefit at all. Such technology is unaffordable. The tenant farmer and sharecropper get into the hands of moneylenders or are thrown off the land their families have farmed for generations so that the landlords can increase their acreage. The mechanized agribusiness is not so labour intensive, so local farmers may be dispossessed, and also unemployed. India has witnessed some particularly ugly scenes.

It is of particular concern to botanists and plant breeders that the wide adoption of the new high-yielding varieties of rice and wheat has caused tremendous change to the genetic resource. The new, uniform varieties are fast replacing the diverse but less productive landraces developed over thousands of years. In Pakistan, for example, all the areas to which access is easy, by road or path, have been planted up with the new cultivars. Yet the variation in the landraces was an essential ingredient in breeding the new crops and will be needed again to improve them and keep up with rapidly evolving pests and diseases. Breeders will

have to work fast to collect the surviving range of types before they are exterminated. This problem is dealt with at greater length in Chapter 11.

Maize

The story of maize reflects, in microcosm, the story of agriculture. It nourished the great civilizations of ancient America and today is the principal food plant of the western world, with more than 700 million tonnes produced annually. Yet it is a monster which cannot live without people. The cob, gift-wrapped in papery layers of husks, does not shatter at maturity and kernels are clamped tightly to the central core. If the cob is not harvested it eventually drops to earth and the seeds, all germinating at once, will choke each other in mass fratricide. Few of the 1,000 or so seedlings would survive to mature and breed, unless people or machines shucked the kernels from the core and sowed them apart.

There are many different varieties of maize, each having its own qualities tailoring it for its environment and end purpose. Most commercial maize today is yellow or white "flint" or "dent". The flint corns are hard, sometimes with a small amount of opaque floury tissue near the centre of the kernel; dent types have a central core of soft starch which shrinks on drying, causing the surface to dent.

Maize arrives at our table as butter-drenched golden cylinders of corn-on-the-cob, sweetcorn out of a tin enlivening many a dreary rice salad, or that party favourite, popcorn. We fry in maize oil and use corn-

flour to thicken gravies and make puddings; we might even take a little as breakfast cereal or drink it as whisky. A huge proportion goes to feed livestock so we eat it eventually, its energy transferred to meat, eggs and dairy products.

In less wealthy parts of the earth, as in Latin America, maize is a vital element of diet, often coarsely ground and made into a paste or gruel. But it is not an ideal food for humans, as it is low in protein. Only recently have plant breeders incorporated genes to improve the nutritional value, and one day these could be of great benefit in preventing malnutrition.

No one knows for sure when or where maize was born, nor its exact ancestry. Very small cobs, only about 20 mm long, found at St Marco's Cave, Tehuacan, Mexico, have been carbon-dated to about 7,000 years old. They bear the hallmark of cultivation; the grains are arranged in several rows on a rachis that does not break up at maturity. Scientists have discovered a two-metre deep accumulation of rubbish and subsequent excavations at this site have revealed a sequence of cobs showing the evolution of maize. Bat Cave in New Mexico concealed an ear of primitive pod popcorn for 5,000 years; complete ears dated at 500 BC are similar to races found in Andean Peru and Bolivia today (though these are distinct from both current or archaeological Mexican maize).

The Indian peoples grew their crops mostly in slash-and-burn clearings in the forest. After three or four years of cultivation the clearings were abandoned for ten years or so until fertility was restored. The Colombian Indians of the Choco region still

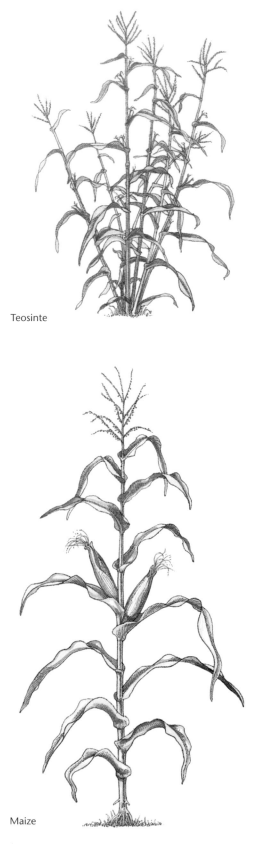

Teosinte

Maize

practise a primitive form of this "swidden" cultivation; they scatter seeds in a freshly cleared plot, only returning to the crop at harvest time. Open grasslands like those of the southern USA and the Argentinian pampas, now major crop producing areas, were not used by early inhabitants because their crude implements could not break the mat of roots. However, in arid areas maize could be grown in sunken fields. In northern Chile, for example, sand was cleared down to a level at which soil moisture was present, then planted. A similar form of agriculture is carried out to this day by the Hopi Indians who dig a series of widely spaced pits, planting about ten kernels in each. They add soil gradually round the plants as they grow.

The centres of maize variability seem linked to the ancient civilizations. The Mexican and lowland Central American dents are associated with the Mayan culture; the variability of maize found in the central Andes is linked with the Incas; the Guatemalan and north Andean flints with the Chibchans. The conical corns of central highland Mexico were the power behind the Aztecs, and Montezuma, the last Emperor, demanded a tribute of some 3,000 bushels each year from his 20 provinces. It is possible that these diverse centres indicate independent origins of maize agriculture.

When Columbus landed on Cuba in 1492, he found maize agriculture in full swing. Indeed maize was being grown right from the mouth of the St Lawrence River in Canada in the north down to central Chile in South America, with some 200 to 300 local varieties. Since then, maize has spread the world over. Cuban and Argentine flint corn gave rise to the true flints used throughout the tropics and subtropics. The flints of temperate zones are often traceable to the northern flints and flour corns of the USA. Dents of Mexican and lowland central America have also spread around the tropics. Some of the Mexican dent corn was taken into the southern USA where in the mid-1800s it crossed naturally with flint

*Maize, **left**, is believed to be derived from the wild grass teosinte, **above left**, its familiar heavy cob being equivalent to teosinte's relatively miniscule ear, **see facing page**. Modern maize, or corn, is a biological monstrosity, created by human cultivation, and quite unable to survive natural conditions. Teosinte can interbreed quite easily with maize; indeed, it has apparently swapped genes with it naturally from ancient times, though breeders have not made much use of it. The explorer C. Lumholz, travelling in Mexico at the turn of the last century, discovered that teosinte was known to be good for the corn. Traditionally seed was saved from one crop for the next year's plantings, and where there was any distance between one cultivation and the next, the isolation could well have led to inbreeding and the associated depression in yield. A dose of strange genes from teosinte would without doubt have injected a healthy spurt of hybrid vigour into the offspring, and rejuvenated the line. The name itself, teosinte, comes from Aztec "teocentli", meaning God's ear of corn. Maybe this reveals the cultural memory of the wild grass being taken into domestication; in many parts of Mexico teosinte is known as "madre de maiz" (mother of maize).*

types. This north-—south fusion gave rise to the dents now of prime commercial importance in the US Corn Belt and which grow so well in southern Europe too.

The early European settlers of the Americas adopted the local varieties of maize with no attempts at breeding better strains until the mid-1800s. Then they started to produce prize corn for exhibition at agricultural fairs, where a standard idealized corn was required. The emphasis was on large cylindrical ears of dent maize produced on single-stem plants with no sideshoots. The craze led unfortunately to the elimination of much variety, and the industry is now based on only two of 200-plus primitive varieties.

Repeatedly crossing one type of maize with itself quickly produces a weakened strain. However, crossing two distinct inbred varieties produces plants far out-yielding the parent types: this is the phenomenon known as heterosis or hybrid vigour, and is important in no crop more than maize. The hybrids are uniform in appearance and all other qualities, which is virtually essential in modern crop production and use. To get a really bumper yield, seed resulting from the cross of two pure parent lines must be planted every time, and cannot be saved from the previous crop, so a specialist producer of hybrid seed is required who maintains the pure lines and produces hybrids for the farmer to sow.

To perform these precise crosses it is essential to ensure that a plant does not fertilize itself. This used to involve a great deal of manual labour in removing the tassels (male flowers) from the plants which were to produce the seed, so that female flowers could be fertilized with pollen only from specified male parent plants. Then in the late 1940s a special form of male sterility that prevented pollen formation was found in some Texas maize. A plant with this factor was suitable for the seed-producing parent of the hybrid and did not require de-tasseling. Its progeny, pollinated by a fertile male line, were self-fertile and had full hybrid vigour.

The Texas male-sterile factor was a rare and lucky find, and thus very widely used. However, this sterility was activated in the cytoplasm of the cells rather than controlled by genes on the chromosomes in the cell nucleus, as such factors usually are.

At one time virtually all commercial maize had the same cytoplasm so though appearing varied, the vast acreages of the crop grown in the USA were identical in this one important feature. The inevitable happened and disease struck in 1970. The fungus *Helminthosporium maydis*, revelling in a warm damp spring, spread like wildfire across the Corn Belt, travelling as much as 150 km a day. At harvest dark clouds of spores billowed around the combines. It was very lucky that this outbreak of southern leaf blight caused no more than 15 percent loss overall, though in some of the southern states more than half the crop was lost. Consider the consequences if the outbreaks had occurred in a country where maize is the primary human staple. In Kenya, for example, where it constitutes more than half of all food consumed, there would have been a tragedy.

Nearly all maize improvement has been achieved by using the genetic resources of the variety within the crop itself. There are

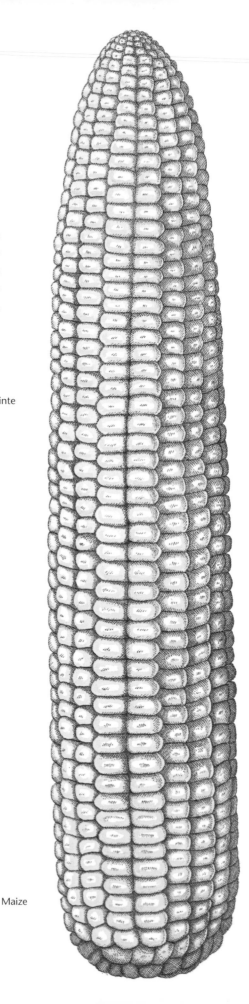

Teosinte

Maize

Many of the primitive maizes show an astonishing range of colour, with the greatest variety of all being found in the central Andes. Various parts of the kernel are white, yellow, lemon, orange, red, purple, brown or pink, and these may also be speckled, spotted, striped or streaked – as illustrated by the assortment, **top left**, *hanging in the house of an Otavalo Indian ready for next year's sowing. These maizes are maintained on innumerable small plots, each with its locally adapted varieties, such as the one illustrated,* **bottom far left**, *in Narino Province, Colombia – a far cry from the huge areas of monoculture in Maryland, USA,* **bottom near left**. *As modern "super-maize" takes over, the genetic diversity preserved in traditional cultivation is all too easily lost.*

just two cultivars into which breeders have introduced genes from *Tripsacum*, a related wild grass, which can be crossed with maize only with difficulty. The local name "teosinte" is used for several wild species. For years one of the wild species, *Zea perennis*, was believed extinct in the wild, kept alive only in botanical gardens. Then in 1977 Rafael Guzman, searching in the remote mountains of southern Mexico, came across a stand of it. Not content with discovering the Holy Grail of maize botanists he then, not far away, came across a stand of a totally new plant, *Zea diploperennis*. Some experts believe it is the true ancestor of maize, and so great is the excitement concerning this wild grass that it has widely been called the botanical find of the century.

Maize is an annual, these two grasses perennial. This conjures dreams of amazing labour-saving fields cropping year after year without resowing. However, making perennials annual is a common feature of domestication because then the plants put an all-out effort into filling the fruits, not wasting food and energy on (to humans) useless vegetative organs and overwintering devices, so making maize perennial is not likely to be a major triumph of Guzman's discoveries.

Zea diploperennis crosses freely with maize, and has many useful features, especially immunity to many of the major virus diseases that ravage corn. It has been estimated that as much as $250 million a year could be saved if only one percent of the US crop were protected.

The tragedy is that these two species only grow in a tiny area of Mexico, but efforts are being made to protect these populations. At

the same time, scientists are investigating methods of transferring desirable qualities, such as disease-resistance and flood tolerance, from them to cultivated maize.

Organic farming

For decades, the richer countries of the world have produced food more and more efficiently, partly through the use of chemicals, such as weedkillers and pesticides, and by the addition of artificial chemical fertilizers. This has caused environmental damage and a reduction in wildlife, and it has not always improved our diet. Whilst there is no doubt that such practices have produced blemish-free vegetables and have generally increased yields, there are negative aspects to such intensive methods, and these are gaining more and more attention, not least in terms of our wish to promote healthier eating habits. Significantly, many chefs are now only using certified organic ingredients in their restaurants, on grounds of flavour as well as health.

We have learned to our cost about the damaging effects of adding pesticides and other chemicals to the environment. Wild animals, especially predatory birds, accumulate chemicals, and can even become infertile, as happened before the pesticide DDT was banned in the early 1970s. Our own bodies also store such substances, and it is not sensible to drench our crops with cocktails of chemicals, many of which are, at least, potentially dangerous.

Recent research has revealed that hazardous, mostly man-made chemicals have contaminated every environment, affecting wildlife, including polar bears, whales, frogs, fish, and even snails, as well as ourselves. As many as 300 man-made chemicals have been found in humans, and it is likely that we are all contaminated in this way, especially in the industrialized countries.

Another unfortunate consequence of mass production and uniformity is the loss of local varieties of, for example, fruits and vegetables. Britain for instance has lost about 2,000 varieties of vegetables in the last 30 or 40 years, and our orchards once grew many more varieties of fruits than they do today. Ironically, with global markets available, countries often find it more economic to import fruits such as apples from the other side of the world, with the transport involved adding to global warming, as well as the practice discouraging home-grown produce.

In Europe, the Common Agricultural Policy has long encouraged intensive farming, which in turn has caused sharp declines in wild birds and a general impoverishment of the agricultural environment. Hedgerows and trees have been removed, and the "cleaner" fields leave fewer weeds and grain for the birds. In the UK for example, tree sparrows have declined by 95 percent, corn buntings by 88 percent and skylarks by 52 percent – danger signals warning us that something is genuinely amiss in the farm-dominated countryside.

What we now see is a global trade in agricultural products, driven by the commercial pressures of big business and the major supermarket chains, and which tends to promote intensive systems. This trend runs counter to the needs of small farmers, and makes it very difficult for them to succeed in producing local varieties for local markets, although we are beginning to see a shift towards this, for example with the success of "farmers' markets".

As with plant conservation, there is a pressing need to encourage local, sustainable systems of crop production, both in developed and in poorer developing countries. While traditional societies have in many cases farmed organically for centuries, it is only in recent years that there has been an upsurge of interest in developed countries.

Organic farming maintains the long-term fertility of the soil and uses fewer inputs, to produce high-quality, nutritious food. The soil fertility is maintained mainly by using crop rotation, often involving legumes that increase the nitrogen content of the soil, composted animal manures, and ground rock minerals. Weeds are controlled by mechanical methods. The use of pesticides, growth regulators, artificial fertilizers, and feed additives is prohibited.

The benefits of organic farming are legion: more success for local farmers; greater plant genetic diversity; fewer chemicals required; and, not least, healthier food. A richer, more heterogeneous environment also reduces the chances of pest epidemics, as does the presence of greater genetic variation in the crops. Intensive systems on the other hand are characterized by large expanses of identical crops, far more susceptible to attack by pathogens and insects, thus creating the vicious circle of requiring chemical protection.

One can even see a situation in which the developed countries will have to learn lessons from traditional more "primitive" farmers, and perhaps it is not too fanciful to imagine a resurgence of smaller, ecologically sensitive farms, encouraged by a general change in attitudes of customers. It is in this context that the organic food lobby can be influential and important.

Britain, for example, has a rapidly expanding market and demand for organic produce, although the government has targeted that only some five percent of farmland should be organic by 2007. It is a hopeful sign that many of the larger supermarkets are now stocking more organic produce, in response to customer demand. About two percent of the US food supply is organic, but sales have increased by some 20 percent annually.

For those concerned about the conservation of the countryside and protection of plants and animals, it is clear that organic agriculture has immense benefits – and that the less intensive methods result in more diverse wildlife.

In Britain, the Soil Association has launched a campaign to persuade the government to aim for a target of 30 percent of the agricultural land and 20 percent of food production to be organic, but despite recent encouraging signs, there is a long way to go before these targets are reached.

Advances in agricultural science have been beneficial, but these need to be carefully integrated into locally relevant sustainable agro-ecosystems, retaining the traditionally close relationship between farmers and their land and time-tested varieties.

The spread of organic agriculture

Key
(land under organic management in hectares)

1	*Australia*	10,000,000
2	*Argentina*	2,960,000
3	*Italy*	1,168,212
4	*USA*	950,000
5	*Brazil*	841,769
6	*Uruguay*	760,000
7	*UK*	724,523
8	*Germany*	696,978
9	*Spain*	665,055
10	*France*	509,000
11	*Canada*	478,700
12	*Bolivia*	364,100

Top 12 countries with land under organic management, either fully converted or in conversion

A massive field of organic spinach, Wellington, Colorado, USA.

Above *In 2004, organic agriculture covered more than 240,000 sq. km in 88 countries – an expanse almost equivalent to that of the UK. The leader was Australia with 10 million ha, followed by Argentina with almost 3 million ha. Since 1996, the UK's organic area has surged from 50,000 to 724,523 ha. Almost 80 percent of UK households buy organic food, spending around £1 billion – more than any other European country apart from Germany. The UK aims for 70 percent of its organic food sales to be home produced by 2010. In total, Europe has 170,000 farms covering 55,000 sq. km. In several nations, including Italy, Sweden, Finland and Switzerland, between 5 and 10 percent of agricultural lands are now organic, while Austria has reached 13 percent.*

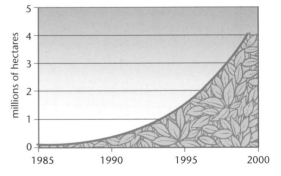

Organic farming in the European Union
In 1999, almost 4 million hectares – 3 percent of the total agricultural land – were devoted to organic farming in the EU. Since 1985, organic farming has been spreading at an average growth rate of 30 percent – more than any other region in the world

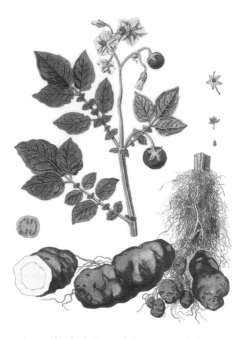

An early depiction of the potato (Solanum tuberosum), **above**, *clearly shows the five-petalled flowers typical of the family Solanaceae. The fruits, too, are typical, resembling those of its relative the tomato (though in the potato's case they are poisonous). Most Andean ancestors of the potato have rather small tubers, but this painting shows quite large ones, though far more irregular than is considered desirable in commercial terms today. Our modern potatoes, conveniently shaped and of lavish size, are nonetheless rather dull vegetables – and will grow only in temperate climates. In Peru, however, many varieties, each with a different flavour, are known, while in the potato's Andean homeland many wild varieties grow, offering hope for cultivation in the tropical lowlands.*

Present problems

Advanced agriculture often causes greed where it should create plenty, and starvation in the wake of apparent munificence. Worse still, it produces new problems as fast as it solves the old. Nature responds with ever more pests and diseases, vastly encouraged by our huge monocultures; mutating as rapidly as the pesticides and fungicides to control them are changed, these attackers are often one jump ahead.

Probably the most notorious crop failure was that of the potato in 1845 and 1846 which caused famine in Ireland and a surge of emigration to the New World. Cultivated for thousands of years in its native Andes, the potato was first brought to Europe around 1570. It did not catch on immediately in its new environment because the early introduction did not prosper in the long-day seasons of temperate regions. But satisfactory clones eventually arose and by

the early nineteenth century the potato had become the major item of diet of the Irish poor, and also of their pigs.

Here was a recipe for disaster: large acreages of one crop built on the slenderest genetic foundation, and the whole peasantry entirely dependent on it. In mid-August 1845 a blight struck the crop on the Isle of Wight and in the course of weeks spread throughout the British Isles. Contemporary accounts state that it destroyed between a quarter and half the crop. The following year the epidemic began earlier and was even more destructive.

It was some time before the nature of the disease was properly understood. In damp conditions the green parts are attacked by the fungus *Phytophthora infestans* so the tubers do not fill out, or if they become infected they are weakened and are rotted by secondary pathogens. By 1926 resistance from related wild species had been bred into the commercial crop and scientists thought they had late blight conquered. Then in 1932 a race of *Phytophthora infestans* arose which overcame the resistance in the new varieties. Eventually a new resistant cultivar was created, which also in due course fell foul of a new race of the disease. This cat and mouse game between breeders and the blight has been going on ever since. Furthermore, this fungus is just one of the many organisms that appear to enjoy the potato as much as we do, and each of which the farmer has to fight tooth and nail.

This example can be repeated with almost every crop plant that has been much altered from its ancestors and is grown on any scale.

The Potato Famine of 1846 was caused by the fungus disease late blight, which swept through the potato fields of Britain, all planted with the same crop variety, in successive waves of destruction. The collapse of the potato harvest produced an enormous social upheaval in the British Isles. In Ireland nine-tenths of the crop rotted; with no produce the peasants were unable to pay their landlords, who went bankrupt. The people either stayed and starved, or emigrated; many went to the New World, such as the Irish immigrants, **left**, *disembarking at New York.*

Pests are even more difficult to control than fungus diseases and are on the whole a worse problem. At least one in ten of the world's million-plus insects is a crop pest, and mice, rats, birds and other creatures are also serious pests. Between them these probably cause the loss of at least a third of world crops. It has been calculated that in India's biggest state, Uttar Pradesh, there are more than 600 million rats, which consume food produced enough to support 100 million people. Breeding can be undertaken to give appreciable resistance to attack by small pests like aphids and eelworms, but not to the larger ones.

This is the sting in the tail of modern agriculture; it is dependent on a vast energy consumption to produce pest and disease controls, fertilizers and weedkillers. High dependence on chemicals seems continually to increase; and in the USA it has been calculated that the proportion of energy expended on these, compared with that on machinery for cultivation and harvesting (itself no mean quantity) is in the region of eight or nine to one.

There is therefore a need for greater use of natural pest controls where these can be encouraged, and of new crop plants that will, at least in the less favoured areas of the earth, crop heavily without the need for such vast inputs of energy.

The future

There is no reason to suggest that our present balance of food crops is the best possible for the future. It clearly places vast strains on the world's economy which may be impossible to maintain. Yet world population increases and people need to be fed.

In the long term one may foresee a world in which the west's present eating habits, concentrated so much on meat, may be changed so that less energy and space are expended on feeding the animals concerned. Another problem is that social structures in many countries result in large numbers of very poor people, and these populations are vulnerable to malnutrition, disease, and even starvation.

But this book is not a forum for these problems. It can only point to other avenues for crop plant development. The scope for change and improvement in present staple crops is still great, and will remain so as long as the variety of types, which are the raw ingredients for modification, survive, as explained in Chapter 11.

Quite a number of other food crops also have scope for improvement – cassava and the various yams for instance both have great possibilities.

Job's tears, *Coix lachryma-jobi*, a grass, is sometimes grown for ornament – the grains being large enough to make beads. It is also a very nutritious cereal which could possibly replace wheat and rice in disease-prone areas. A few local strains exist but no big effort has ever been made to develop it. The Australian grass *Echinochloa turnerana*, too, has potential; it can yield crops of good

The pummelo (Citrus maxima), **above**, *is still an unfamiliar fruit in western eyes. A native of tropical South-east Asia, it is a large-fruited, sweet-flavoured species which has often been used in hybridization with sweet oranges and mandarins. The resulting offspring are large and of good flavour, but unfortunately most have large numbers of seeds. Recently it has been found that some varieties of pummelo are salt-tolerant, opening up possibilities of cultivating citrus fruits on saline soils. The forests of South-east Asia have many hundreds of wild fruit-tree species, and around 125 are locally cultivated. Some offer potentially wider markets, such as the mangosteen* (Garsinia mangostana), *the rambutan* (Nephelium lappaceum), *or even the durian* (Durio zibethinus) – *with its delightful taste but infamous smell.*

grain after one thorough initial watering, but has never been domesticated.

Breadfruit, from a tree, has a very high starch content and is made into flour for bread, or the fermented pulp into dough cakes. Many varieties, as in bananas, are seedless. The related jackfruit, weighing 20 kg or sometimes much more, is boiled and roasted, or eaten fresh for dessert when ripe; and a less important relation, the champedak, likewise. These *Artocarpus* species originated between India and Malaysia, but breadfruit and jackfruit are quite widely grown in the tropics as subsistence crops. They are permanent plants rather than annual ones so, once established, need little attention; they have considerable potential as food plants.

Most agriculturalists will be reluctant to change the overall pattern of cultivation in the world's fertile areas at present, though there would seem to be advantages in using other crops than the big three cereals in some circumstances, which may be desirable especially where pests and diseases overtake breeders' achievements. Where we can most profitably develop seems to be in using land which has been considered basically unusable for food crops – not, one hopes, by eradicating existing wild plants where these are desirable, but in dry semi-desert and possibly eroded areas.

Most particularly new possibilities exist for saline soils, which in some areas are unfortunately the result of over-salination by irrigation schemes. At least a quarter of once-used cropland has been ruined in this way. Estimates indicate that about 48 million hectares of the world's irrigated land has

Inflorescence of amaranth. This was once a very important crop in Central America, with a history of 7,000 years of cultivation, and is still important in Peru where it is grown mainly on mountain slopes. Each seedhead produces more than 100,000 tiny seeds, rich in protein. Amaranths are also grown in South-east Asia, where the leaves are eaten as vegetables.

Illustration of an Aztec man harvesting amaranth (Amaranthus sp.) from the Florentine Codex, a sixteenth-century work illustrating Historia Verdadera de la Historia de Nueva España, *an encyclopedic account of Aztec society written by Fray Bernardino de Sahagun. The illustration is reproduced from Francisco del Paso y Troncoso's colour edition of the Florentine Codex 1905–07.*

been affected by salt accumulation, and that this is growing at about two million hectares per year. In many places underground water reserves that are saline could be pumped up to irrigate suitable crops.

There are some edible wild plants that grow naturally on salt and could certainly be grown as crops, having good yields and food value: these include pickleweeds, saltworts, saltbushes and saltgrasses. The latter have been tried in an alkaline, dried out lakebed in Mexico which now supports livestock. One of the few flowering plants that grow in the sea, eel grass, is harvested by the Seri Indians of coastal Mexico who grind the grain into flour. This marine grass offers the surprising prospect of using the seas to grow our bread.

Then there are wild relatives of numerous currently grown crops which tolerate salt and could be bred with existing strains to create new varieties for saline soils. Among these are a sugar beet and several cereals – barley, millets, sorghum, rice and wheat. A form of garden beet is known to grow in 75 percent sea water, and there are tolerant forms of date palm, pistachio and pummelo. New hybrid tomatoes have been bred which can grow in 70 percent sea water, derived from garden varieties crossed with a small-fruited species found on the shores of the Galapagos Islands.

Plants with potential in really arid areas include several yams from the Australian deserts: used by the Aborigines, they have scarcely been investigated. The Aborigines' intimate knowledge of food plants in drylands remains a largely untapped resource. Another intriguing root is the buffalo gourd (*Cucurbita foetidissima*) from the most arid places in Mexico and the south-western United States. Its vast starch-filled tuber – three to four metres long – can weigh 145 kg in old specimens, and up to 30 kg after only two seasons. On top of this it produces round fruits whose seeds contain 30–35 percent protein and up to 34 percent of oil. A hectare of this plant could produce two tonnes of seed, yielding nearly a tonne of vegetable oil. Once established, care needed is minimal; the potential crop value approaches 700 dollars per hectare. Wild populations of the buffalo gourd show considerable genetic variation – some plants are prolific, some have larger fruits, some are barren – so there is much scope for the plant breeder to select new strains from the wild.

The Somalian yeheb nut (*Cordeauxia edulis*), recently on the verge of extinction, is now being cultivated in East Africa for its nutritious and tasty nuts, while its foliage can provide animal fodder in its arid haunts.

Another dry-loving plant is the morama bean (*Bauhinia esculenta*) of the Kalahari Desert, where some Bushmen live for months on end on little else. Old plants produce bean-bearing runners in a 25-metre radius, though six metres is more normal. The beans, high in both protein and oil content, are usually toasted, or pounded into a flour from which a soup or porridge is made. As well as the nutritious beans it has tuberous roots which make a good vegetable when eaten young, at one or two kg. Old roots can reach 100 kg or more. When watered under cultivation, the morama bean grows very quickly and produces a phenomenal crop.

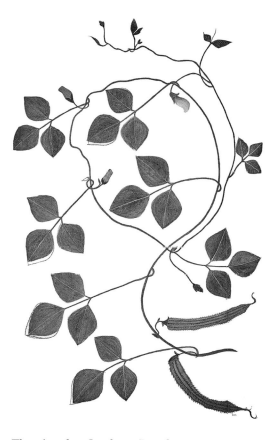

The winged or Goa bean (Psophocarpus tetragonolobus) *has been grown for many generations as a "poor man's crop" in South-east Asia and Papua New Guinea, but only recently has the remarkable versatility and nutritional value of this food attracted international attention. Every part of the plant can be eaten. The immature pod, considered a delicacy, is the plant's most popular part. But it is the dried seed which offers hope against malnutrition from protein and vitamin deficiency in the tropics. It provides high-quality protein, and substantial quantities of oil. In many respects it is a rival to the soybean, which is difficult to cultivate in the humid tropics. The winged bean is now being grown in more than 70 countries, and is a strong contender for further development.*

Twists of the cyanobacterium Spirulina, **above left**, *each just a quarter of a millimetre long, are potentially of great importance in feeding a hungry world.* Spirulina *is suitable for industrial production in open ponds or closed polythene tubes, and out-performs all of the major crops with yields ten times higher than wheat. A mat of* Spirulina *is shown,* **above right**, *on a conveyor belt at a processing plant in Mexico.*

Under more normal humid, hot conditions the wax gourd (*Benincasa hispida*) grows more easily than its relations the squashes, pumpkins and melons, and so fast that three or four crops could be raised in a year. The gourds are two metres long and up to 35 kg in weight, and their crisp and juicy white flesh can be eaten at any stage of growth, in various ways – it has a mild flavour, and is prepared as a cooked vegetable, in soups, or as a sweet mixed with syrup. Their thick waxy skin allows the gourds to be stored for up to a year without refrigeration.

A final example of a number of barely appreciated possibilities is the winged bean (*Psophocarpus tetragonolobus*). Like the morama bean it has multiple uses – it is really a dream of a crop, producing edible leaves, flowers, seedpods, seeds and tubers. Food researchers have dubbed the plant "a supermarket on a stalk". The leaves taste like spinach, the sautéed flowers like mushrooms; the young pods are like green beans, the young seeds like peas; the tubers, richer in protein than potato, yam or cassava, can be boiled, fried, baked or roasted. The mature seeds are like soybeans,

and also yield oil. Like most legumes it harbours nitrogen-fixing bacteria in root nodules, so improves the soil in which it grows. This powerfully climbing bean originated in South-east Asia and is already an important garden plant in several tropical countries. Many hundreds of wild varieties in South-east Asia are now being studied; researchers hope to find short-stemmed, sturdy characters suitable for extensive cultivation. Perhaps the most extraordinary aspect of the winged bean story is that this important food plant should have been ignored for so long.

Seaweeds were quite widely eaten in Europe and North America within this century, though seldom today. The Hawaiians, Indonesians and other Pacific dwellers eat them in some quantity, but it is in China and notably Japan that they are most prized. Hundreds of thousands of tonnes are collected in Japanese waters annually for home consumption and export to China. Improved methods of collection – by cutting rather than grappling – will indeed be needed if the seaweeds are not to be eradicated here. In both countries there is some cultivation of seaweeds, especially kelps.

Seaweeds are rich in sugars and starches, have some protein, and many contain vitamins A and C. They also contain iodine and, partly for this reason, the iodine-deficiency disease goitre is almost unknown in Japan.

An even less expected food has been made from the primitive cyanobacterium *Spirulina*, found for example on Lake Texcoco, Mexico, and the small lakes surrounding Lake Chad in Africa. These are very salt and have high alkalinity content, in the region of pH 11, and virtually nothing but *Spirulina* will tolerate these conditions.

The pre-conquest Aztecs collected the algal mats that drifted in the wind to the lake edges, and made sun-dried biscuits of them, as do the shore dwellers of Lake Chad today. These biscuits have the phenomenal food value of 70 percent protein dry weight; they contain vitamins, especially B12, and are readily digestible. The health food market was not slow on the uptake and *Spirulina* pills are now on sale, while the alga is being produced commercially in Mexico as a high-protein, high-carotene additive to chicken-feed. It is this carotene which gives the African flamingos their glorious pinkness. Wider use as a human protein supplement, maybe a meat substitute like the soybean analogues, is within the bounds of possibility, though at the present time its production cost is high.

This cyanobacterium consists of single cells, sometimes arranged in chains or groups, and is rich in chlorophyll. There are other micro-organisms that are rather similar, though without chlorophyll, which are being developed more extensively. The great advantage of these is that they can be made in factories in relatively small areas rather than cultivated in vast fields like the staple cereals. As such they may help us to save further areas of natural vegetation.

To recapitulate, we have seen how vulnerable our crops are. Over the years we have come to depend on a smaller and smaller number of crop species. Today, fewer than 20 crops supply more than 80 percent of the world's needs; three (wheat, rice and maize) provide more than half. And within each of these crops, the variation is diminishing: more than half Canada's prairie wheatland is planted with a single variety. Their uniformity has made them far more vulnerable to epidemics of pests and disease which, because each cultivar is planted over such enormous areas, can be catastrophic. In 1970 the US maize crop was devastated by southern leaf blight because 80 percent of the plants were susceptible to the disease.

Plant breeders fight a constant battle to keep one step ahead of pests and diseases. Resistant races of insects, fungi and bacteria rapidly develop to overcome resistance in uniform crop species. For every barley variety marketed in Britain in the last 20 years, resistance to mildew has been lost within three years.

The answer is, as so often, to be found in the wild. Recent exploration in the Himalayan foothills of north-eastern India has yielded many primitive rice cultivars resistant to major diseases and pests, including blast, bacterial blight, tungro virus, gall midge and stem borer. A single population of wild rice *(Oryza nivara)* from central India is the only known source of resistance to grassy stunt virus, which was once a serious disease. Resistance from this rare wild plant was bred into a cultivar, IR36, which has since become the most widely grown variety of rice – but for how long?

Plant breeders are able to transfer favourable characteristics from one species or cultivar to another. They find these characteristics from among the diversity of the crop itself as well as from the species related to it. Yet it is the advanced cultivars which, by their very success, are wiping out the variable and diverse plants on which future breeding will rely: the success of the agricultural revolution is destroying these resources essential to the future of good farming. This must surely be one of the most worrying trends in the world today and one which demands the attention of every politician and policy-maker.

There seems to be no need that the plant kingdom cannot supply. There is no challenge to which it cannot rise. All that is required is a measure of human ingenuity, intelligence, and the political will to care for it.

CHAPTER FIVE

The Spice of Life

There are a few things in life we cannot do without: sufficient food, good water and protection from cold. Then there are those luxuries that – for the fortunate – add to the joy of living: a scented bath, a gourmet meal, a fine old malt whisky.

Luxuries are, by definition, rarities. They are also subject to the whim of fashion. A local product, gathered from the wild, or casually cultivated by an indigenous community, suddenly becomes a craze in another culture often halfway across the world – who had heard of jojoba oil cosmetics a few years ago? The initial burst of enthusiasm is the danger point for a wild plant, when collection is heavy but before it becomes profitable to cultivate the species. The sandalwood craze of the seventeenth century nearly wiped out the species. Today wild palm hearts are a delicacy that threatens other trees with extinction.

Every country has its larder of wild produce, a cornucopia for the initiated. Pre-agricultural people exploited hundreds of plants for food, but with the adoption of agriculture there was less time for gathering wild plants, and less need, so much knowledge faded. However, when the maize crop failed in Tanzania in 1952, some people managed to support themselves in the famine conditions on wild produce. Their relative health surprised government officers since only baobab fruit had been officially recognized as a wild food. Subsequent research identified 40 bush foods, with an additional ten roots that could be eaten in times of dire emergency.

In places like Britain hardly any people rely on what they can glean from the

Rajasthani women spread chilli peppers on the ground to dry in the sun.

hedgerows. Wild foods are regarded as luxuries – a few fashionable restaurants serve nettle soup in season. In much of Europe, however, especially the south, country people use native plants for food and medicine as an important part of their life. For many other peoples, the knowledge of the hunter-gatherer is still a necessity. The native people of the Kalahari Desert in Africa, Australia, the Philippines, India, and South America all have complex diets in which plant foods are of critical importance.

Another example can be seen in the ethnic minority communities of Yunnan Province in south-west China, and especially in the tropical south, bordering Myanmar and Laos. Here the local people have long lived in close harmony with nature, using the forests and other habitats as sources of food, firewood, building materials and medicines, and harvesting these sustainably.

Of all the plants gathered from the wild, fungi are among the most fascinating: the autumnal fungus foray is a special event in Russia and northern Europe, with the exception of Britain. One of the most luxurious food items in the west is an underground fungus, the Perigord (or black) truffle *(Tuber melanosporum)*, which can cost up to $1,000 (£535) per pound ($2,220/£1,190 per kg). It resembles a piece of coke – blackish, warty and up to 7 cm across. It is gathered using pigs or specially trained dogs that rootle around the forest floors, notably in the Dordogne region in France, to find it by its smell. Other fungus species, such as the Italian white truffle or the morel, cèpe and chanterelle, not quite so highly prized, are widespread and commonly eaten.

Fragaria chiloensis

The modern cultivated strawberry, Fragaria ananassa, *is the offspring of a union between two American species brought independently to Europe:* F. virginiana *(native to eastern North America) and* F. chiloensis, **above** *(native to Chile, and locally cultivated in western North America). The first hybrid seems to have been an accident, arising in a garden near Amsterdam around 1750. The European wild strawberry,* Fragaria vesca, **below**, *was probably the species cultivated by the Romans in 200 BC.*

Fragaria vesca

Luxury fruit and vegetables

In terms of luxury some fruits are classier than others. For instance, in Britain the pawpaw, persimmon, lychee and loquat are regarded as something out of the ordinary, a distinction they lack in their countries of origin. The luxury value thus appears to be based on "foreignness" and unavailability, especially as the very weight and perishability of fruit make them expensive to transport and handle, while the opening up of export markets may shift the price of fruits such as mangoes beyond the reach of the average local pocket.

A few fruits transcend the requirement for foreignness, the strawberry being an obvious example. There are about 35 different species of strawberry found wild in northern temperate areas, into the subtropics and in South America. All bear edible fruit and were probably items of fare in the hunter-gatherer's diet. Certainly the Romans cultivated a strawberry, probably the European one, way back in 200 BC, and this species was also grown in medieval Europe. Today's strawberries are one of the most important soft fruit crops, with well over a million tonnes produced each year, sometimes in poor countries as cash crops for export.

Vegetable luxuries are fewer. With the admonitions of nutritionists ringing in our ears we eat up greens because they are good for us. Only a few are regarded as delicacies – the tender tips of asparagus shoots and the thistle-like heads of artichokes, for instance, are dignified as luxuries by their usual position of appetizers

at meals. Indeed, artichokes were the most esteemed vegetable in England during the seventeenth century.

Asparagus developed from wild European species. Grown in beds of rich soil supplied annually with manure, it is an extravagant vegetable since it takes several years to become fully established, only crops over about six weeks in the year, yet remains in place for long periods.

As for the artichoke, only the fleshy bases of the bracts and the base of the flower are eaten, while the aptly named choke – the immature flower – is discarded: a sophisticated approach indeed. The tender central leaf stalks are also edible. (Cardoon, a close relative of artichoke, is occasionally cultivated for the leaf stalks, which are blanched before eating.)

It is remarkable that anyone discovered it was possible to eat true thistles. Yet they can be delicious – not the initial taste on biting, but the meltingly sweet aftertaste. Even the heavily armed marsh thistle has an edible central core, and stalks or young shoots of many other species have been eaten as a delicacy since hunter-gatherer times, wherever they are found growing in the northern hemisphere.

Palm hearts are an increasingly popular luxury vegetable – crisp, succulent, and delicately flavoured. Much eaten locally for centuries, they are now being canned and exported. However, their popularity is of special concern to conservationists because in harvesting the single heart – the cabbage-like bud which is the only growing point – the tree is killed. One canful of only a dozen segments requires the death of a tree at least

The milk thistle (Silybum marianum), **left**, *a wild plant of the Mediterranean, is generally edible, including the leaves when trimmed of their prickles. In Turkey people crunch the stems raw, though the delicate-flavoured young shoots are more usually boiled.*

The artichoke (Cynara scolymus), **above**, *is an ancient crop native to the Mediterranean and Canary Isles. It was cultivated by both the Greeks and the Romans.*

eight years old and up to 20 metres high. The major sources are wild Brazilian trees of various species including *Acrocomia*, *Euterpe*, *Prestoea*, *Roystonea* and *Sabal*. Only recently have plantations been established to replace wild sources, reputedly in part of Brazil's nature reserves. Nine South American countries have palm heart (palmito) industries based on *Euterpe* palms, mostly aimed at the export market. This industry is however being investigated for its potential to support extractive reserves from which the product could be harvested sustainably, thus maintaining the forest cover.

Food flavourings – herbs and spices

A way of creating a good meal, if one cannot afford to start with expensive ingredients, is to enliven humble fare with flavourings. Herbs and spices are widely used today to add variety and interest to the diet. But in medieval times, an era before refrigeration, their principal function was to preserve food or disguise its flavour.

There is a strong link between herbs as food flavourings and herbs as medicines, and this is not coincidental. Often the medically active substances are precisely the flavourings the cook will wish to incorporate into a dish. Chopped garlic releases a sulphur-containing volatile oil which has antibiotic activity, is an expectorant, a weak vermifuge and also helps to lower blood pressure. Spearmint contains the volatile oil carvone which stimulates gastric secretions,

so is useful for stomach disturbances, while thyme contains a good antiseptic. In his *English Hus-Wife* of 1615, Gervase Markham recommends a mixture of pennyroyal, coriander seeds, aniseed, sweet fennel seeds, caraway seeds and brandy as a cure for digestive problems.

These herbs are all of European (mainly Mediterranean) origin, and were doubtless found in the domestic herb gardens of seventeenth-century England. But even at this time, trade in exotic spices from the Orient was significant too (it had, of course, been in progress for several thousand years). There was a great deal of money involved. The Mediterranean cities of Alexandria, Genoa and Venice owed their prosperity to the spice trade, and its history is woven into that of the peoples and powers along the great trading lines. The caravan routes led westward from China, India, Sri Lanka and Indonesia, crossing the River Indus at Attock,

Spice plants

Saffron comes from Crocus sativus, **right,** *an attractive flower native to Greece, but now grown widely in warmer non-tropical regions. The part used is the three-branched orange-red style which must be hand-picked from the newly opened flowers. Up to a quarter of a million flowers are necessary to produce a pound of saffron, but only a tiny amount is needed to impart the flavour and yellow colour to a dish.*

Vanilla, **below right,** *is the product of an American orchid* Vanilla planifolia. *The Aztecs used it to flavour chocolate. The long, yellow seed pods are picked unripe and have to be cured before the flavour develops.*

Cardamom (Elettaria cardamomum), **below,** *is related to ginger. Before 1800 all cardomoms were wild-collected from the jungles of southern India, but since then plantations have been established. Care is needed in gathering, as the little seeds must remain within the capsules if the flavour is to be retained; thus they are hand-cut with scissors before they are fully ripe, and carefully dried.*

Black pepper is provided by the berries of the Indian jungle vine Piper nigrum, **below.** *Always an important spice, pepper was a luxury in medieval times; but like salt, it is now an almost universal condiment. The plant was originally called Poivrea, in honour of Pierre Poivre, the adventurous eighteenth-century French spice hunter.*

Piper nigrum

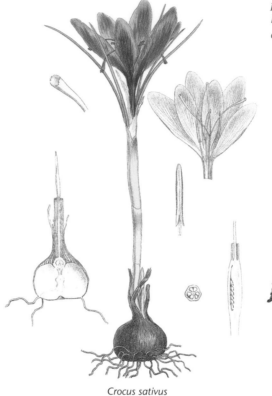

Crocus sativus

Elettaria cardamomum

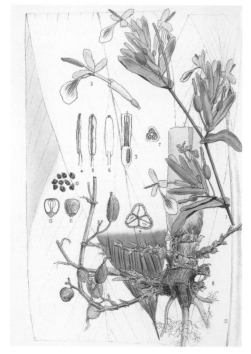

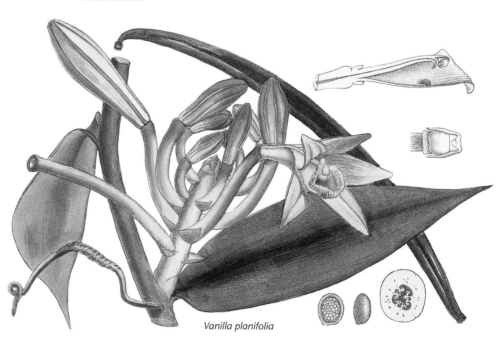

Vanilla planifolia

Chocolate comes from the cacao tree (Theobroma cacao), **above**, *which originates in north-eastern South America. Its centre of diversity is north-east Peru and adjacent Ecuador, areas where much of the forest is threatened.*

on through Peshawar and over the Khyber Pass, through Afghanistan and Iran to Babylon, then onward to Europe. Most Europeans thought of spices as "Arabian" – an ignorance encouraged by the traders.

One of the most important spices was black pepper *(Piper nigrum)*. It was known to the Romans, and in medieval times rents were sometimes paid with it – hence the term peppercorn rent. This now means a nominal sum; but in times past pepper was a scarce luxury, and it was sometimes the preferred currency. Botanically unrelated are the New World peppers (mostly *Capsicum annuum*); Jamaican pepper *(Pimenta dioica*, also known as pimento or allspice) belongs to the myrtle family.

The capsicums belong to the same family as potato. *Capsicum annuum* takes many forms, from the sweet bell pepper which is delicious as a major part of a meal (and which, when powdered, is paprika) to hot chilli pepper which must be used sparingly. Chilli peppers range in size from the bell, big as a fist, to the fiery little chiltepin, no bigger than a pea. Most of the ornamental peppers grown for their colourful fruits and sold as houseplants also belong to this species. The Mexicans were using peppers, probably similar to the chiltepin, as early as 7000 BC.

Some families are particularly rich in flavourings. The Umbelliferae (carrot family) provides angelica, aniseed, caraway, celery, chervil, coriander, cumin, dill, fennel, lovage, parsley, rock samphire and sweet Cicely. The Labiatae (mint family) includes balm, basil, marjoram, the various mints, rosemary, sage, savories, and thyme. From the Compositae (daisy family) we have alecost, chamomile,

southernwood, tansy, tarragon and wormwood. Others – bay, capers, cardamoms, chives, cinnamon, citrus, cloves, fenugreek, garlic, ginger, horseradish, mace, mustard, nutmeg, saffron and turmeric – are scattered through the plant kingdom and represent almost every part of a plant. The most expensive spices are saffron (iris family), vanilla (orchid family) and cardamom (ginger family).

A few spices are still gathered from the wild. Melegueta pepper *(Aframomum melegueta)*, which is a substitute for the expensive cardamom, is generally collected wild in West Africa. Allspice is collected wild in Central America. Wild-collected herbs and spices are often regarded as inferior since their flavouring is not so intense. Unlike the other food plants in which breeding has generally reduced the strength of flavours, these plants when cultivated often have increased volatile oil contents, and are generally preferred to wild sources.

Chocolate, coffee and tea

In the America of the pre-Aztecs, chocolate was the "Food of the Gods". The Aztecs too valued cocoa, not only as a drink but also as a means of paying taxes. In fact cocoa beans were used as currency until 1887. It was not until the sixteenth century that cocoa and its products became known in Europe. The Aztec emperor Montezuma is thought to have entertained the Spanish conquistador Hernando Cortez by offering him "xocoatl" – a drink prepared from cocoa beans and flavoured with capsicum peppers. The result must have been rather bitter. It

Tea is made from the young leaves of the bushy Camellia sinensis, **above**. *It varies in flavour immensely depending on the variety of plant, and method of production: in China, green tea, made from fresh, dried leaves, is commonest; black (red) tea is made from part-fermented leaves.*

was apparently Spanish nuns who first mixed chocolate with vanilla and sugar, making it palatable to European taste. Today, two-thirds of the world's cocoa is grown in Africa, notably in Ghana. The Portuguese took cocoa first to Príncipe Island and São Tomé, and thence to the African mainland.

Coffee, in contrast, comes mostly from a plant that grows wild in the forests of the Ethiopian massif, *Coffea arabica*. Red coffee berries develop from beautiful white jasmine-scented flowers on laurel-like bushes. The berries, much loved by monkeys which raid the plantations given the chance, each contain two seeds generally known as beans. In the fifteenth century it was discovered that chewing coffee leaves and beans helped to relieve pain, hunger and fatigue – uses long known to the indigenous people. Even today, dried coffee berries are eaten by East Africans, or fermented to produce an alcoholic drink. As a hot beverage coffee only became popular after its introduction to Europe; by 1675 there were already 3,000 coffee houses in London, and by the 1960s coffee had become the largest single item (in value) imported by the United States. It is widely grown in the tropics, and is the most important crop in many Latin American countries. Another species, *Coffea canephora,* produces "robusta" coffee; it is grown mainly in South-east Asia, Central and West Africa, and in Brazil.

Tea, another beverage containing the stimulant caffeine, originated from a different continent – maybe in Tibet, or possibly further north in central Asia. Tea

(Camellia sinensis) grows wild mainly in south-west China and north-east India, and there are two main varieties: *C. sinensis* var. *sinensis* (China tea), and *C. sinensis assamica* (Assam tea). It has been cultivated for more than 4,000 years in China, where its earliest uses were probably medicinal. Commercial tea did not reach Europe until after coffee drinking had become well established. Habits too vary from place to place: delicate teas are often drunk without milk or sugar, sometimes with lemon; in Russia tea is sometimes taken with jam, while in Tibet rancid yak butter is added. Now there are more than 3,000 different types and blends of tea, exported all over the world, and more than three million tonnes are produced from over 25 countries, with over 80 percent coming from India, China, Sri Lanka, Indonesia and East Africa (especially Kenya).

Many other teas or tisanes are made by infusing dried leaves, like maté made from *Ilex paraguariensis*, a Paraguayan holly. Some people prefer this to tea or coffee since it contains almost no tannin. Mint tea is favoured in North Africa and the Middle East, and many herbs like chamomile are made into herbal teas. One is endangered even: in the Sierra Nevada of Spain, tea fanatics search for the elusive and endemic *Artemisia granatensis*. In addition to making tea from its leaves and flowers, they also use it to prepare an alcoholic drink.

Today most commercial production of cocoa, coffee and tea is in places far removed from the plants' original home. Yet these places may contain wild relatives vital to plant breeders. For example, wild and semi-wild cacao trees from the upper

Amazon Basin have already increased yields and provided resistance against several diseases. The genetic base of the world's cocoa crop is still very narrow, and today new sources of disease-resistance are badly needed. However, these forests, especially in Peru and Ecuador, are fast coming under threat from resettlement schemes, logging and drilling for oil.

The danger of relying on a small gene pool is also illustrated by coffee. (Most of Brazil's "typica" coffee originates from the progeny of just one tree, introduced from East Africa via the Caribbean.) In the 1870s, coffee rust *(Hemileia vastatrix)* devastated the Sri Lankan crop, and the disease destroyed most of Brazil's coffee in 1970. Finding resistance to this disease is a top priority. Wild and semi-wild coffees, mainly from the highlands of south-west Ethiopia, provide a rich source of genetic diversity, and their conservation is a priority. Some of the wild strains are resistant to rust disease. But many of these montane forests have been damaged or destroyed for timber, fuelwood and to create agricultural land, and only about 770 sq. miles now remain.

Booze and tipples

Virtually any carbohydrate can provide a food from which yeasts can create alcohol – the process of fermentation. Thus the sugars in grapes produce wine, the starches in barley, Scotch whisky; apples, rice, bananas, cherries, potatoes – any good sugar or starch source will do. Even the sap of some palms can be used, such as the rare Chilean wine palm *(Jubaea chilensis)*.

Vitis vinifera

The grape vine
Most European grapes are grown on rootstocks from a combination of three North American wild vines, because virtually all the European vines were destroyed between 1870 and 1900 by Phylloxera, *an insect pest. Resistant American plants thus saved the industry.*

The popular Mexican drink mescal, and the more refined tequila, result from distilling the juice of Agave tequilana. *The centres of the plants are first chopped up and baked in stone-lined pits,* **above**, *then crushed in (traditionally donkey-powered) circular roller presses, before distillation. Another popular* Agave *drink is pulque, made from the sap that fills the plant's centre after the flower-bud has been cut out.*

The flavour depends on the type of plant used. It is the female flowers of the hop *(Humulus lupulus)* which give beer its distinctive bitter flavour. Three wild American ancestors of modern hops are the essential basis of the crop grown in almost half the world's hoplands. In England alone it is reckoned that these wild plants have contributed some £8 million a year. Of course, the flavour of a good beer does not just depend on the hops and other ingredients: the ageing process and the way the beer is stored are important too, as any real ale enthusiast knows.

Making beer or wine sounds very simple, but it is actually rather difficult to produce a good alcoholic drink with a pleasant flavour. It is important that only the desired organism is introduced; villains such as *Mycoderma aceti*, the vinegar bug, are ever ready to take advantage of unsterile conditions.

One of the less enticing alcoholic potions is surely chicha. (There are also non-alcoholic versions.) This is made in South America by fermentation of maize, manioc, palm fruits, plantains, potatoes and so on. The starting material is chewed and spat out so that the enzymes in saliva catalyse the conversion. Kava is the Polynesian equivalent, made from chewed roots of *Piper methysticum* or *Macropiper latfolium*.

For those who like a stronger tipple, spirits are distilled from the products of fermentation. They can have up to 50 percent alcohol content. Plants are again used as the starting point. Whiskies are made from barley or rye, gin from juniper berries, and rum from molasses (a product of sugar cane). Vodka originated during the fourteenth century in Russia, where potatoes are frequently used as the raw material for fermentation.

Even some of the plants used for making wine and spirits may come under threat: it may be too late to save one plant that was extensively used in wine making – *Pseudophoenix ekmanii*, a palm from the Dominican Republic. Sadly it is now endangered or possibly extinct in the wild.

Drugs and stimulants

More important than the hallucinatory drugs illustrated opposite are opium derived from the opium poppy *(Papaver somniferum* ssp. *somniferum)* and cocaine derived from coca *(Erythroxylum coca)*; because of their value in medicine they are described in Chapter 7. The effect of potentially hallucinogenic plants depends on the amount of the active ingredient taken, though it is often impossible to administer precise doses because of the natural variations from plant to plant. The recipient's state of mind and what they expect to happen are very influential as well. A timid person who dared to take cannabis in sufficient quantity, whether smoked in a cigarette or eaten in a sweetmeat, may end up paralysed with fear; yet it emboldened Al-Hasan's infamous hashashins to become the holy terrorists of Islam in the twelfth century. Furthermore, the critical dose will vary between individuals; the amount that will leave one person unmoved may pole-axe another.

Inebriants and the less potent caffeine-containing stimulants (tea, coffee and cocoa) are all drunk; the daily stimulant in the Far East and Southern Asia, used by

Hallucinogenic plants

About 60 species of plants worldwide, from South and Central America especially, have hallucinatory properties, allowing participants to have visions and mystical experiences, and shamans to commune with spirits, heal, and direct people in rituals.

Many morning glory relatives are hallucinatory – the seeds contain lysergic acid derivatives, and were used by the Aztecs in divine rites. Seeds of garden morning glories (Ipomoea tricolor), **above***, have been sampled by modern "hippies"; however, they produce very nasty hangovers including nausea, constipation and vertigo.*

Peyote – the "mescal button" (a spherical, spineless cactus, Lophophora williamsii), **above** *– and several other cacti of Mexico and southern USA are ritually consumed to provoke kaleidoscopic coloured visions, either wonderful or horrible, and ritual dancing. (Eating them must take some willpower as they are nauseating.) Shamans contact the spirit world under their influence and the peyote cult is often involved with Christian ritual. The ease with which dried "buttons" can be kept, and the vivid picture of their effects given in books such as Carlos Castaneda's Teachings of Don Juan has caused the cult to spread widely in North America.*

Cannabis sativa, **above,** *has been known to the Chinese for over 4,000 years, and probably almost as long to the Indians. Marijuana, kif, grass, hemp – it has many names – was used to produce euphoria, hallucinations and, in Africa, for communal ritual. Cannabis resin, or hashish, taken in large quantities increases aggressiveness and courage – the Persian word "ashishin" gives rise to both "assassin" and "hashish". Cannabis is now very widely taken by westerners and is widely, if dubiously, regarded as no more harmful than alcohol, though there is obvious evidence of the damage caused by both.*

Amanita muscaria

Psilocybe semilanceata

The fly agaric (Amanita muscaria, the "fairy" toadstool), **top***, was taken communally by tribesmen in Siberia to produce euphoria, a desire to dance, and happy visual hallucinations. Poorer Tartars waited till participants in the ritual needed to pass water, and drank their urine with equally satisfactory effects. It was used, too, in priestly ritual in India, where the fungus was worshipped.*

Smaller mushrooms such as certain species of Psilocybe, **above***, were ritually used in Central America and now often supplement Christian rituals.*

Left *Dark projecting grains in these cereal heads are infections of the fungus ergot. In the Middle Ages those who ate bread contaminated with ergot were seized with "St Anthony's Fire", suffering gangrene and possible loss of limbs, convulsions and sometimes violent hallucinations either of being pursued by wild beasts, or of a religious nature. Chemical extraction of lysergic acid, a constituent of ergot, led to the creation of the drug LSD which psychiatrists have used in therapy, while individuals use it for "instant Zen". But it is a very dangerous hallucinatory, leading for instance to an often fatal belief that one can fly, and/or to permanent psychosis.*

millions of people, is however chewed. This is betel, which contains the mild alkaloid arecoline. The betel nut, part of the fruit of the betel palm *(Areca catechu)*, is grated, mixed with lime and spices and wrapped in a leaf of *Piper betel,* a relative of black pepper. The nut can be used fresh or cured, though cured it contains less tannin, and therefore tastes better. The alkaloid is used medically to cure tapeworm infections, and chewing betel has this useful side effect. Countering this benefit, it encourages heavy spitting, while long-term use discolours the teeth (in some islands of the Pacific, black teeth are a mark of high social standing); and recently it has been suspected that the lime encourages cancer.

Tobacco is one of the most controversial and widely used drugs. There are huge campaigns to stop people smoking on the grounds of individual health as well as the passive smoking inflicted on others. These are countered by the devotees who claim their health is their own business and that the do-gooders ought to set about removing pollution from other sources such as car exhausts before interfering with personal freedom. Ironically, revenues derived from taxes on tobacco are an important part of the national budget in some countries.

The main tobacco species is *Nicotiana tabacum,* and there are several cultivars and types of tobacco, whose characteristics are also affected by cultivation and climate. The most important are flue-cured tobacco (named after the metal flues that distribute heat in the curing barns), a thin-leaved, light coloured tobacco known as "burley", oriental tobacco, and air-cured tobacco.

These have specific uses in cigars, cigarettes, pipe and chewing tobacco, and snuff, though they are sometimes blended.

The chief active ingredient is the alkaloid nicotine which tranquillizes the nervous system. It is manufactured in the roots and transported to the leaves, the part harvested. Tobacco is now grown in many countries, from North, Central and South America to Europe, China and India. Global tobacco production is expected to reach over seven million tonnes by 2010. Most of this ends up smoked for human enjoyment, but nicotine's other important use is as an insecticide (also from *N. rustica*): it is lethal to most insects and, although in principle it is toxic to mammals, in low concentrations is reasonably safe.

There are two species of tobacco plant in domestication, Nicotiana tabacum, **right**, *and* N. rustica, *both South American members of the potato family. The earliest record of use is a bas-relief from a temple at Palenque in Mexico, which has been dated to* AD *432. It shows a Mayan priest blowing smoke through a tubular pipe during a ceremony. The smoking habit was taken up enthusiastically by the Spanish settlers early in the sixteenth century, and rapidly spread across Europe and into many parts of Asia. Even today many peoples are newly adopting tobacco, as the tranquil backwaters of the world fall under western influence.*

Perfumes and cosmetics

As a means of survival, the human sense of smell is relatively unimportant. We do not sniff out our mates, proclaim territory, or find our way home with it. We can just about locate where dinner is being cooked, or a dead mouse under the floorboards, or tell if the house is on fire. Yet, for all its uselessness, few luxuries compare with the sheer pleasure of a beautiful perfume.

Just as medicine and cookery have a lot in common, so do cookery and perfumery. The same kinds of substances are active in all three fields. Subtleties of taste in food are virtually all smell. The taste buds just below the surface of the tongue are merely sensitive to saltiness, sourness, sweetness and

bitterness; all flavours are actually detected by the olfactory cells in the mucous membrane of the nose.

Perfumes have a long history. Part of the mummification of rich Egyptians involved filling the body cavity with myrrh, cassia and other perfumes after the intestines had been removed. The Queen of Sheba visited King Solomon to negotiate sole rights for the overland trade in frankincense and other goods between southern Arabia and the rich markets of the north. In these times, as early as 3,000 years ago, all people in the Fertile Crescent – even labourers – used cosmetics and perfumes, the latter usually based on sesame oil or on wine. The Romans learned about cosmetics from the Egyptians and used them widely.

Emphasis on scent was again strong in Europe from the Dark Ages onwards. This may have been in part due to the generally malodorous conditions that prevailed, especially in town houses. Pestilence often resulted, in which posies of fragrant flowers were carried to ward off disease. The nursery rhyme "Ring a Roses" recounts the course of bubonic plague in gruesome detail; the pocket full of posies was sadly ineffectual.

Hygiene seems to have been practically non-existent in western Europe at least until the sixteenth century. Scent must have been highly valued, and indeed for many generations the strewing of aromatic herbs was practised. Plants such as sweet woodruff, sweet rush, hyssop, various mints, germander and chamomile were spread on the floors to release their fragrance when trodden on, periodically to be swept up and renewed.

Queen Elizabeth I had a female retainer who had to provide perfumed plants for the household. Apart from alleviating bad smells, these plants deterred fleas, lice and other vermin. They were also placed among clothing for both fragrance and vermicidal qualities. (The present day visitor to Nepal will find pungent artemisias still being used for this purpose.) The court of Louis XV of France was known as "la Cour Parfumée" because of the extravagant use of scent. Napoleon Bonaparte is said to have had 60 bottles of double-extract jasmine delivered every month, and two quarts of violet cologne were supplied to him each week.

The perfume industry has been revolutionized this century by the creations of organic chemists, but even now expensive perfumes rely on natural products, with the synthetics adding new notes and tones to the perfumers' repertoire. The most important natural oils in perfumery are absolutes derived from the flowers of rose, jasmine, and orange. Oil of orange is pressed from orange peel, and lemon and bergamot oil also come from citrus fruit peels. Labdanum and patchouli are derived from leaves; cinnamon bark is used in perfumery as well as in cookery; the lichen oakmoss contributes earthy notes; cedarwood a refreshing resinous scent.

The current trend in cosmetics, too, is to emphasize their herbal origins, thus we have apple shampoos and avocado face packs, but the ingredients have changed little over millennia. Such poisonous excesses as ceruse (white lead) and belladonna are no longer in use, but the basic oils, gums, starches, lamp-

Flowers are used extensively in perfumery. Unlike the other essential oils produced by plants, flower perfumes are designed to attract pollinating insects. They are derived from, and replace, pigments akin to chlorophyll. Thus brightly coloured flowers tend to lack perfume, whereas highly scented flowers, such as ylang ylang (Cananga odorata), and tuberose (Polyanthes tuberosa), **above,** *are pale.*

Lily of the valley (Convallaria), **above,** *is an important scent in perfumery; it has thick, fleshy petals which hold the strong perfume. In this it resembles honeysuckle (Lonicera) which is pollinated by night-flying moths and is particularly heavily scented. Night-pollinated blooms are typically pale-coloured, glowing like soft beacon lights at dusk.*

black and colourings keep on turning up with accompanying fanfares of advertising in bright new disguises.

Even so, many women today follow a similar make-up routine to that of Egyptians 3,000 or more years ago. In the time of Tutankhamun a lady of consequence would prepare her face for the day by first applying a cleansing cream and natron (a sort of natural washing soda, similar to bath crystals), then pluck her eyebrows. She would outline her eyes with bold green sweeps of powdered malachite in oil, and touch up her eyebrows with lampblack. Red ochre was rubbed into the lips and cheeks. Henna was used on the nails, the palms of the hands and soles of the feet to tint them a delicate pink.

Where clothes are few or lacking, body paint becomes necessary as a mark of caste or tribe. Some Native Africans use mineral colourings such as red or yellow ochre mixed with oils. Ancient British warriors stained themselves with woad, and Native Americans used natural plant dyes such as madder and indigo, which were semi-permanent and did not smudge. Sometimes the plant dyes were made even more fast with aluminium and iron salts, mordants well known in the dyeing industry.

The ancient craft of dyeing is based on plant lore. The wrappings from mummies have revealed that the Egyptians were familiar with the dye indigo 6,000 years ago. Indigo was produced from a number of plants including European woad (Isatis tinctoria), the Chinese indigo (Polygonum tinctorium), and – chief source of supply before synthetics – the shrub Indigofera tinctoria.

Madder, produced by several plants, was also known to the Egyptians and remained an important plant dye until World War II. The colouring principle, alizarin, gives reds and oranges on mordanting. This industrial use of plants was suspended when, in 1856, an 18-year-old chemist, W. H. Perkin, created mauveine by the oxidation of crude aniline, thus producing the first synthetic dyestuff. Others followed rapidly; in this industry, as in others, chemists took naturally occurring substances for their pattern. Notably, alizarin was manufactured in 1869 from the coal-tar product anthracene, the first vegetable dyestuff to be made artificially, and artificial indigo was created in 1878, although natural indigoes are still widely used in non-industrial situations.

Among other plants, dyes can be obtained from weld (Reseda luteola), saffron (Crocus sativus), fustic (Maclura tinctoria), onion skins, logwood chips and lichens. Craft-minded people nowadays like to use these natural dyes to colour hand-spun, hand-woven cloth. Thus luxuries come full circle; the brilliant dyes of industrial origin are rejected by those who prefer a return to nature. Will this resurgent interest in wild products, from herb teas to plant dyes, from flower-based soaps and make-up to rare spices and food, be converted into yet another threat to our harassed wild environments – whether or not we, as consumers, intended this? There are signs that this may be happening.

Very different plants may release similar odours. The popular perfume of sweet violet (Viola odorata), **above,** *is closely resembled by scents from the roots of costus (Aucklandia costus), the grass vetivert or "khus-khus" (Vetiveria zizanioïdes), and orris (Iris pallida). Roots of vetivert are woven into sun-blinds in India and Malaysia, to be moistened in the heat of the day and thus both cooling and scenting rooms.*

The perfume of sandalwood is characteristic of the heavy scents of the Orient. Attar of sandalwood (from chips of the wood) is used extensively, and vast quantities of scented red sandalwood timber are used to build temples and houses, and in religious caskets and ornaments. This has led to the severe decline of several Santalum *species, including* Santalum album, **above.** *One species (S. fernandezianum), from Juan Fernandez, the Robinson Crusoe island, is now extinct.*

CHAPTER SIX

Green Wealth

If money makes the world go around, then plants are the engine. Fortunes are made and lost by people dealing in commodities they never see, from timber to rattan, from soya to coffee, from oil seeds to resin, rubber and latex. They buy and sell as prices rise and fall, gambling not only on this year's produce, but on "futures" too. In every field of human enterprise, plants are the primary producers: energy, industry, clothing, housing, crafts and fine arts – all of them depend on plants.

Food crops are big money, whether bought and sold within their countries of origin, or entering international trade. Many foods – from peanuts to luxury strawberries or artichokes – are grown entirely for overseas sale; they are "cash crops" designed for export earnings. This is fine where the country growing them has also enough capacity to grow its own staple needs; or where, like Israel, it can import staples in exchange for high-value cash exports. But in poor countries the cash crop market, which encourages farmers to plant non-essential foods, can lead to disaster: if the buyer switches to another source, the market drops, or disease strikes, the farmers have no money to buy the food they could have grown.

The greatest "green wealth" is the wild itself. Perhaps if its unexplored potential were assigned a dollar value in practical terms we would be less hasty to clear it for short-sighted reasons. Perhaps if the industries that stand to benefit from new plant discoveries paid "royalties" on the plants they borrow, the countries concerned would have adequate funds to conserve their heritage.

Today is a particularly exciting and challenging period for botanists. As oil becomes dearer and the end of the oil era comes into sight, possibly only 30 to 50 years away, industry is turning to our inheritance of plant capabilities for new solutions – plants to replace fossil fuels and all their derivations: plastics and oils, insecticides, chemicals, waxes, gums, pectins and resins. Their source lies in the wild, particularly the tropical jungles, and in the knowledge of wild plants and cultivars held by local peoples. The marine flora, too, has secrets of immense value.

Nowhere is the disaster of wilderness destruction, deforestation and pollution more evident than in this: that potential benefactors, still awaiting our attention, may be gone before we can test and cultivate them. Fortunes are stillborn, industries of the future denied.

Energy from plants

Petroleum, coal and natural gas are all fossil fuels – solar energy stored by photosynthesizing plants laid down in the Carboniferous era, about 300 million years ago. These are not renewable reserves: we now consume more coal every year than accumulated in 10,000 years. The hideous irony is that we are using the products of this primeval sunshine to destroy today's green wealth – it drives the tractors, chainsaws, water pumps, mining machinery and so on that are so swiftly eradicating natural habitats.

Replacement energy supplies are under urgent investigation: tidal flow, wind and wave power – and plants. Many countries

US wheatfields, seen here in Washington State, cover much of the areas that once supported natural prairie grassland. Cereal production in North America is big business, often organized on a vast scale.

Brazil is growing sugar cane to convert by fermentation into alcohol for fuel; its National Alcohol Programme is being scrutinized the world over. Begun in 1975, production reached 12.4 billion litres in 1988, by which time 90 percent of cars produced in Brazil had engines modified for this fuel. Cars running on alcohol produce far less pollution than those using petrol. However, tracts of Amazon forest have been destroyed for the gasohol farms, like the one shown **left**.

Eventually Brazil intends that cassava should replace sugar cane as the main gasohol feedstock.

Brazilian "Alcool" stations offer the motorist a choice of fuel grades, **left.** *Standard car engines run on a blend of up to 15 percent alcohol with little or no tuning, and can be modified to use up to 40 percent; new cars have been designed that will run on alcohol alone.*

The main by-product of gasohol production is a very rich organic soup which is sometimes used in fertilizer or cattle feed manufacture. It is occasionally dumped in rivers, where it is a serious water pollutant and has wiped out many fish stocks.

are now considering, or even implementing, "biomass" production programmes: fermenting and distilling the energy-rich products of present day photosynthesis, to obtain fuel alcohol or "gasohol". Developing countries, in particular, hope to achieve a degree of self-sufficiency in energy production. However, there is concern over growing crops for fuel, rather than food, where malnutrition is so widespread.

The raw material for biomass production varies from country to country. Maize is the major distillery feedstock in the USA. Sweet sorghum is also being considered, as are wastes from cheese making, citrus and food processing industries. Beets are favoured in New Zealand. In South-east Asia, sweet potato seems the best crop since it matures more quickly than sugar cane or cassava, the other main contenders.

In India, Pakistan, Sudan and various African nations, molasses has potential. In the past, it has been regarded as a waste product of the sugar cane industry (because of lack of storage and transport facilities), and returned to the land. Today it is considered the major feedstock for alcohol production programmes.

Energy from the forests

In the long term, the most economic form of plant life for fuel alcohol production is wood. Yields per hectare are very high, there is more total biomass available from trees than from any other source; and agricultural inputs needed are much lower than for many crops. The biggest disadvantages are that trees take a long time to start

producing; and that lignocellulose, their major component, is not so easily processed as starch or sugars. Nonetheless, many people now envisage eucalyptus on a five-year coppice rotation as the "oil well" of the future. There are plans to grow it in Florida on old mining sites. Canada too has a policy of taking energy from the forest, using its native softwoods.

Wood has, of course, been our primary fuel source throughout most of history. In large parts of the world, this is still the case: more than a third of the human race depends on wood alone for cooking and heating; half the timber cut every year is for fuel, most of it used in developing nations. Worldwide, biomass (mostly wood) provides about 30 percent of the total energy supply in developing countries, and in some countries, such as Nepal and some African states, woodfuels provide 80 percent.

Large energy plantations are planned, or already operational, in many parts of the world, especially in developing countries where firewood is vital. For millions of people, meeting everyday needs is increasingly difficult as the forests disappear: in parts of Africa so many trees have gone in the last two decades that people (often women and children) have to walk many miles a day for firewood; the landscape is reduced to semi-barrenness. Where fuelwood is so scarce that it must be bought, heating the supper pot can cost as much as filling it.

One answer is to plant woodlots in gardens, marketplaces, along roadsides, as shelter belts, as counter-erosion measures, even seeded from the air on to wastelands. Improving the efficiency of fuel use is

The obvious way to provide an adequate supply of firewood is to plant trees specifically for fuel, especially in areas now denuded. In planning fuelwood plantations, several important questions must be answered. How quickly do the trees grow? Species that quickly produce branches of one or two centimetres diameter may be ideal for small cooking stoves. Is it a good dense wood that burns well? Will it grow on poor soils or steep slopes? Is it a firewood acceptable under local customs? Those trees most likely to prove useful in fuelwood plantations are pioneer species which naturally colonize deforested areas. They withstand poor soil conditions and many are legumes with nitrogen-fixing root nodules.

This stand of fast-growing Leucaena leucocephala, **left**, planted for fuel near Poona, India, is only 3–4 years old. Already it is supplying local people with firewood, for domestic use (**above**).

important too. An open fire, the traditional way of cooking in many countries, wastes 90 percent of heat. Even fairly basic stoves can improve on this; providing energy-efficient stoves cheaply can help a great deal.

About three-quarters of tropical land is not really suited to conventional agriculture (though 35 percent of the tropical population lives on it); here the best possible agricultural use of land may be tree production. It is conservative of resources, guards against erosion and flooding, and can even improve poor quality soil.

There are already some heartening success stories of firewood production. The beautiful flowering shrub *Calliandra calothyrsus*, transported to Indonesia from Guatemala in 1936, has been adopted as a firewood crop. Plantations now cover 30,000 hectares in Java. The dense wood burns well and is a good size for domestic cooking, though too small for lumber. In South Korea the leguminous shrubs *Lespedeza bicolor* and *L. thunbergii* bind and enrich the bare soil, provide fuelwood and foliage for livestock, and act as a nurse crop around pine seedlings to protect the tender shoots. The Indian neem (*Azadirachta indica*) tree, introduced to the Accra plains of Ghana early this century, keeps up with the local demand for wood, and is reseeded by bats which eat the fruit. Neem is also widely naturalized on the coast of East Africa and is being increasingly used for carving as a replacement for the now endangered native hardwoods such as muhugu ("mahogany" *Brachylaena huillensis*) and African blackwood (*Dalbergia melanoxylon*). Above all, the fast-growing Australian eucalypts are very

*Wood is not the only fuel fit for burning. In eastern central Africa, the tall marsh plant papyrus (*Cyperus papyrus*) from which the Egyptians once made paper, is now a fuel candidate. Able to produce an impressive biomass of 32 tonnes per hectare per year, papyrus is one of the most productive plants in the world. It is not dense enough to burn well on its own, but machinery used for making briquettes from waste straw and wood chippings has been modified to turn out compressed papyrus logs which burn well.*

To make paper, papyrus pith was cut into thin slices which were laid down both vertically and horizontally. They were then pressed, dried in the sun, and finally smoothed. This craft was certainly practised as early as 1500 BC in Egypt.

promising, producing a hard, resinous, fiercely-burning wood which grows well in many parts of the world.

Fuelwood is also an important source of income. Kenya exports charcoal to the Arabian Gulf, and Surinam ships charcoal to northern Europe. On a local scale, too, the production and sale of wood can be an important component of the economy: for instance it has been estimated that at least 6,000 families support themselves by supplying essential fuelwood to the city of Maputo in Mozambique.

Green oil

Another way of using plants for fuel is to extract the oils they produce and use these as a direct replacement for petroleum. For example, a 3:1 mixture of sunflower oil and diesel is similar in efficiency to straight diesel. This would come as no surprise to Rudolph Diesel, the inventor, who wrote as long ago as 1911, "The diesel engine can be fed with vegetable oils and would help considerably in the development of agriculture of the countries which will use it." Plant oils can also be used as "extenders" to make petroleum-based fuels go further. And plant oils in industry could become increasingly important as raw materials for the chemical industry. One consideration, however, is whether land is best used in this way: if more fuel is used (in tractors, sprays, extracting equipment etc.) to produce these oil crops than they finally yield, there would clearly be no point at all in growing them. The energy ratio (comparing the amount produced with the amount used) tends to

*Drought-tolerant jojoba helps the greening of the desert in California, **top**, while providing a valuable crop for the farmer, **above**. Jojoba has been found to fill the vacuum left by sperm whale oil, banned by the USA in 1970; animal conservationists and industrialists alike are delighted that an acceptable substitute has been found for this endangered mammal's precious oil.*

be favourable, however: values of between 3:1 and 10:1 have been estimated.

The crops that can produce fuel oil include olive, sunflower, peanut, soybean, palms, rape, sesame, castor oil and eucalyptus, and some exotic species are now being investigated for potential production in less favoured areas. This variety of possibilities is fortunate, since it means that a species can be chosen to suit its environment, not pressed into service in conditions where it cannot thrive. The Brazilian babassu palm *(Attalea speciosa)* for instance could be grown on impoverished savannahs worldwide. It can produce 60 tonnes of nuts per hectare; on dry distillation these would yield nine tonnes of methyl alcohol, nine tonnes of high-grade charcoal, plus edible oil and cattle cake from the kernels, and pigfood from the pulp. Oil from babassu seeds is estimated to be worth five times as much to the Brazilian economy as its coffee crop. Pure stands of babassu now

cover about 200,000 sq. km, north from the Amazon Basin.

The petroleum nut *(Pittosporum resiniferum)* comes from the Philippines; its nuts yield a petroleum-like oil which, even when green, burns brightly when lit. Recently, this species has been planted in the Philippines and stands can produce 3,000 litres per hectare. The gopher plant *(Euphorbia biglandulosa)* produces a latex which can be broken down into petroleum derivatives. This biennial needs very little water so can occupy desert areas where food crops cannot be grown. Copaiba *(Copaifera spp.)* is an Amazonian tree which the locals tap for oil. It is used medicinally, and in varnish manufacture, and the oil can be substituted for diesel. It is also in demand as an aromatherapy oil, especially in Japan.

Leading the new oil hopes is jojoba *(Simmondsia chinensis)*, the oil of which has gained renown as an ingredient of shampoos; it comes from the fruit of a bush that grows wild in the Sonoran Desert of Mexico and part of south-western USA. Its seeds contain up to 60 percent of a light yellow, odourless, liquid wax. Though nuts have long been collected from the wild, it was only recognized as being of potential importance in the late 1960s and first planted out on a field scale in the mid-1970s. Cultivation has now spread to arid lands in many places including the USA and Mexico, Israel, Sudan, the Middle East, parts of South-east Asia and Australia. Jojoba oil has a clinging quality which makes it an essential engine lubricant; it is even as good as sperm whale oil for nourishing high-quality leather goods; and it is an ideal cosmetic foundation.

It needs limited irrigation in its first year and very little thereafter because a tap root may penetrate 30 metres. It tolerates alkaline and saline soils and is claimed to stop desert land degradation. Jojoba starts bearing seeds at three years old and reaches maximum productivity at around ten years. It is now cultivated in many arid or semi-arid regions, including Australia, Argentina, Israel and Peru, with commercial plantations covering some 8,000 hectares. Recently the seeds have been found to contain chemical compounds that suppress the appetite, and these are being developed for human use, as an aid to slimming.

Paints, polishes and perfection

Art and many facets of modern life such as luxury confections or fashion owe much to plant gums. The finer grades are used in clarifying liqueurs, finishing silk, and preparing artists' watercolours; lesser grades in pharmaceuticals, confectionery, printing inks, dyeing, and as a size for fibres; and the poorest in inks, emulsions and matches. Gums are also used extensively in cosmetics. Famous among these natural gums are gum arabic, gum tragacanth, manila copal and kauri copal. But the most astonishing is guar *(Cyamopsis tetragonolobus)*, grown in Pakistan and the southern USA – it has applications in the food, paper, cosmetics, pharmaceutical and tobacco industries and is also used in the oil industry as a stabilizer for drilling lines. Frankincense *(Boswellia* spp.*)* and myrrh *(Commiphora* spp.*)*, of Biblical fame, are still used in incense and perfumes.

Resins have an extraordinary number of uses: in varnishes, polishes, and floor coverings; printing inks, sizes and waterproofing compounds; soap and cement. From resin, is made oil or spirit of turpentine, an important solvent; distillations of resin go into medicines and ointments. Pines are a notable source of resin; in some areas, they even flavour wine, notably the retsina wines of Greece.

One of the finest plant waxes is secreted from the young leaves of the Brazilian wax palm *(Copernicia prunifera)*. Though expensive, it is the main ingredient of car and furniture polishes, and also goes into varnishes and cosmetics. It grows mainly in north-east Brazil and northern Argentina and Paraguay. Another promising wax plant is the tall herb *Calathea lutea*, known as cauassú. A common riverside plant and colonizer of waste ground, it could readily be developed into a valuable cottage industry in its native Amazonia.

The world on wheels

Rubber is another vital substance for humanity. Synthetics cannot match its qualities; ordinary motor tyres need at least 20 percent natural rubber, and tyres for heavy equipment and aircraft virtually 100 percent. The most familiar source is *Hevea brasiliensis*. This tropical tree from the Amazon jungles has latex vessels just under the bark which yield a milky fluid when cut. The tree is ready for its first tapping at five years old and should be productive for at least 30 years. The best selected clones produce two tonnes of dry rubber per hectare each year.

The Brazilian plantations once created fortunes, but fell foul of South American leaf blight and never recovered properly. It was a great piece of luck that the seeds smuggled out of Brazil to Kew in 1876 (p. 48) were not carrying this destructive disease, and that the Old World plantations that resulted have remained free of it. The earliest commercial rubber plantations in the east were in Sri Lanka and Malaysia, although nowadays the main producing countries are Thailand, Indonesia and India.

Latex is in fact found in many plants. The Belgians conquered the Congo in 1883 and aimed to produce rubber from the inferior *Landolphia* vines. During World War II a

Guayule (Parthenium argentatum) *is a latex-bearing plant from the deserts of Mexico and Texas. Its importance as a major supplier of natural rubber fluctuates with market forces, but it is of special interest because it can be grown on desert land.*

A wild rubber tree is tapped, **below,** *for "caucho" in traditional manner in the rainforest of Madre de Dios, Peru. European explorers of Brazil and Peru recognized the economic potential of rubber when they learned its waterproofing qualities from forest Indians.*

dandelion, *Taraxacum koksaghyz*, supplied some of Russia's rubber needs. Latex from species of *Castilla, Manihot, Hancornia, Cryptostegia* and others has also been exploited commercially on a local scale and some of these might replace *Hevea* if the Far Eastern plantations ever become subject to serious disease. Wild stands of guayule *(Parthenium argentatum)*, native to the southern USA and Mexico, have been exploited for latex since the early twentieth century: in 1910 guayule supplied ten percent of the world's natural rubber – 50 percent of the USA's. When, during World War II, supplies of rubber from South-east Asia were cut off, the government invested $30 million in an emergency programme to revive guayule, and it was grown on 11,000 hectares of prime land in California. When trade ways were reopened, these fields were burned; but seeds of the 25 varieties that had been selected were stored in the gene bank of Fort Collins in Colorado. The World Bank has now forecast that demand for rubber will outstrip supply within the next ten years, so attention is again focused on guayule.

Pest control – the natural way

Plants, being incapable of active movement, have evolved many ways of deterring animal predators. Perhaps the most fascinating – and potentially useful – are the many chemicals they contain that are distasteful or poisonous to insects and can thus be used as insecticides. For example, pyrethrin, extracted from *Chrysanthemum cinerariifolium*, is a poison more deadly to insects

than DDT, yet hardly toxic to mammals or plants; it is non-persistent too, and so environmentally safe. Pyrethrin can be safely used as an insecticide in the house where food is processed and stored. Unlike the synthetic DDT, wonder-worker in its time – especially against the malaria-carrying mosquitoes – it does not get into the animal food chain. The plants are now grown widely in uplands in Rwanda, Kenya, Tanzania, Ecuador and Japan.

Nicotine and derris are also killers of insect pests. *Derris*, a genus of tropical woody climbers, kills beetles and weevils as well, but unlike nicotine, has the advantage of being harmless to mammals.

Some plants offer alternatives to immediate death in controlling insect pests. *Ajuga remota*, a member of the mint family, suppresses their appetites; while tannins from plants like sugar maple harm the insects' digestive systems, thus rendering them vulnerable to predators.

Certain species of *Tagetes*, ancestors of French and African marigolds, release into the soil a sulphur compound that can destroy nematode worms (eelworms) within a radius of a metre. Early South American farmers kept eelworms at bay from their potatoes by interplanting *Tagetes*.

Other plants disturb insect life cycles by producing "chromenes" – chemicals that interfere with the hormones of juvenile or larval forms and so prevent the insect going through the life stages that lead to a breeding adult. Such are the Indian neem tree *(Azadirachta indica)*, a useful fuelwood and medicinal tree, and its relation the Persian lilac *(Melia azedarach)*. Their extract

is distasteful to insects in the first place and, if they persist with it, prevents regular moults and kills them. One of the advantages of the often-planted neem is that its pest-deterrent extract can be made by local farmers. Wild birds have even been noticed incorporating neem in their nests, presumably to help reduce parasite infestations. Neem has so many uses, ranging from fuel, timber, woodcarving and shade, to producing soap, toothpaste and lotions, and medicines for treating skin disorders and arthritis, that it has been dubbed the "tree for solving global problems".

An astonishing adaptation occurs in a wild potato from Bolivia, *Solanum berthaultii*: the plant is covered with tiny sticky hairs which entrap the sap-sucking aphids that try to attack it, and the leaves give off a warning odour or "pheromone" which is identical to that produced by the aphids themselves if alarmed. A few cultivated varieties also produce the pheromone alarm signal, but in inadequate quantities. It should be possible to breed this character (and possibly also the hairiness) into cultivars in adequate aphid-repellent amounts, thus making a great deal of insecticide redundant. About 5.9 percent of the US potato crop, worth $1.8 billion per year, is lost to insects and about 22 percent to diseases, many of them transmitted by insect pests. The percentage is probably greater elsewhere. So the "Scary Hairy Wild Potato" (as it has become known) could be one of the botanical finds of the century.

Feeding the soil

There are numerous ways of improving soil fertility without using artificial fertilizers. One of the best is digging in farmyard manure – dung and urine from stock mixed with soiled straw or some other form of litter. This adds nutrients, especially the three most important – nitrogen, potassium and phosphorus – and improves soil quality, opening up heavy soils and giving body to light ones.

Since herbivorous animals feed off plants, their excreta is a plant product, and their droppings should be a means of enriching soil the world over. Alas, in many countries, fuelwood is so scarce that animal droppings are dried and burned. As one travels in poorer parts of Turkey, India, and other eastern lands in summer one will see people fashioning cow dung into "cakes" which are laid out in the sun to dry for fuel. In India alone, between 60 and 80 million tonnes (300–400 million tonnes fresh weight) are burned every year. In richer countries, too, less animal manure is used as mixed farming gives way to monoculture.

Dead or uprooted plants can be composted to form a valuable soil improver, as many gardeners and small farmers know. The advantage of composting is that the unwanted remains of crop plants are not wasted. Cereal stubbles used to be ploughed back into the soil, but now they are often burned in the interests of "efficiency" – an extraordinary waste of resources in an energy-hungry age, as well as a source of aerial pollution. Many countries however have now banned stubble-burning, or applied strict controls.

The hyacinth bean (Dolichos lablab) *is a legume with considerable future potential. The young pods have many culinary uses, while the foliage can provide hay and silage. Some medicinal applications are recorded too. Probably native to Asia, the hyacinth bean or "Lablab", as it is sometimes called, is now grown widely in India as green manure, it has been introduced to Africa, and spread generally in the tropics where it readily becomes naturalized. Several varieties are known, including a Caribbean climbing form in Trinidad.*

Another way of improving soil fertility with plants is green manuring. A green manure is a crop that is grown only to be ploughed back into the soil without harvesting. It is usually planted between major crops, and conserves soil nutrients. Various plants are used. In Britain, mustard is grown on chalk, lupins on sandy soils, tares on heavy soils, and Italian ryegrass in various circumstances.

Of these, tares and lupins are leguminous, that is, they are members of the pea family that form root nodules in which nitrogen-fixing bacteria dwell, to the mutual advantage of both organisms.

By interplanting with a legume it is possible to improve soil fertility even for a growing crop. Beans and maize can be interplanted to considerable benefit, as farmers discovered very early. The arrangement cuts down weeds, and slows the spread of insects and disease; the beans fix nitrogen, which improves soil fertility, and the maize provides support for the beans. Maize yields are unaltered, and bean yields are only slightly lower, so in effect you get the beans for free.

Fixation of atmospheric nitrogen is extremely important in flooded rice paddies. The floating aquatic fern *Azolla* is one prime source: the cyanobacterium *Anabaena azolla* lives in association with it, and can fix as much as three kilogrammes of atmospheric nitrogen per hectare per day, which is released into the soil when the fern dies. This is particularly important in China and Vietnam – in Vietnam there is a temple dedicated to *Azolla*. Lastly, in the soil and in and among the roots and basal shoots are various nitrogen-fixing bacteria. The presence of these organisms has helped to keep rice yields traditionally as high as one or two tonnes per hectare, without added fertilizer.

Wood – resource under threat

Wood's uses are legion, its potential almost inexhaustible. For all our new technology, it is still the most extensively traded wild product in the world. The vast evergreen forests of the north, both natural and planted, provide us with softwoods – conifers such as pines, spruces, larches, fir and cedar. Often having just one species covering a wide area, they are relatively easy to exploit. The natural vegetation of many warmer regions, by contrast, is predominantly hardwood forest – trees such as oak, sycamore, ash, eucalyptus, teak and mahogany. These are very mixed communities, which makes extraction that much more difficult and expensive. In tropical rainforest there may be only a few species of commercial importance per hectare. Conifers are preferred by most large-scale wood-using industries because they are light, easy to handle and work, and generally speaking are rather similar – so that one species can be substituted for another without problem. The tropical hardwoods, on the other hand, have long been prized for their beauty and durability; like the temperate hardwoods, especially beech, they are valued for furniture, veneers, carvings and luxury goods.

Export of timber from developing countries is on the increase. South-east Asia now accounts for three-quarters of international trade in tropical hardwoods. The bulk of timber exports from developing countries are as logs, sawn wood and panels; only about eight percent is paper and pulp. This demand for wood and paper, mainly in industrialized countries, in one of the main pressures driving global deforestation. The USA, Europe and Japan consume more than half of the world's timber, and almost three-quarters of the world's paper (mainly prepared from trees). We all need to encourage wood and paper suppliers to source timber responsibly, from replaceable, sustainable stocks.

Paper is an essential civilized commodity, as vital to hygiene as to literacy, and places one of the major demands on the forest. The main source of paper fibre today is wood pulp; demand could not be met by the old materials – fibres of grasses like esparto and bamboo, of kenaf, or the most famous source, from which paper gets its name, the papyrus rush of Egypt. In basic papermaking the wood is ground under water, in such a way as to separate the fibres without damaging them. The watery mass is then drained on a moving wire mesh belt, leaving the fibres meshed together in a thin sheet. This is dried and smoothed over heated rollers, to emerge as a continuous roll. For better quality paper, a chemical and pressurized digestion process is used; if this is taken further, the cellulose in the wood dissolves entirely to produce viscose, from which the very fine thread rayon is made.

Virtually every part of a tree can be used for paper and board making – trunk, branches, leaves, roots, even charred wood after forest fires.

The world uses an enormous amount of paper, cardboard and allied packaging materials. Although we live in an age of electronic communication, global demand for paper continues to rise. American newspapers are notoriously large, and one bumper issue of the *New York Times* is the end product of 400 hectares of forest.

Indonesia, for example, is being deforested at two million hectares per year, mainly to supply pulp and paper, and most of the trees come from clear-felling native tropical forests, destroying ecosystems and also impacting on the lives of local forest-dwelling people. From Siberia and Canada to the tropics, destructive logging continues apace, although efforts are being made to produce wood and pulp from sustainable plantations, involving species as diverse as eucalypts, pines, spruces, and the tropical *Leucaena leucocephala* and *Gmelina arborea*. Eucalypts are particularly suitable, as they grow fast and are easily planted and harvested. Brazil now grows nearly 300,000 hectares of eucalypts for pulp, mainly for export.

Much of this forest, of course, is grown specially for making paper; in northern countries, particularly, it accounts for a formidable area of unnaturally covered land. However, paper pulp continues to be obtained from natural forest, by giant machines designed to remove every last tree and shrub. Some of the devastated, eroded land in both Java and Papua New Guinea has been created in this manner from virgin rainforest.

We need paper, but there are alternatives to wood pulp which must be worth pursuing, if only to stop destruction of

Craftsman working in a woodcarving co-operative in Malindi on the coast of Kenya. Here more than 500 trained and highly skilled carvers produce beautiful pieces, mainly to sell to tourists as souvenirs. The whole process, from raw logs to finished articles, takes place on this one site.

The Kenyan carving industry has up to 80,000 carvers, with some 400,000 dependants. Livelihoods are under threat due to dwindling hardwood resources; about 50,000 trees per year are carved, contributing to the degradation of globally important, biodiversity-rich East African coastal forests.

The favoured wood used to be "mahogany" (Brachylaena huillensis) and African blackwood (Dalbergia melanoxylon), species which are now threatened. The carvers are switching to use "good woods", such as the common introduced species neem (Azadirachta indica). Use of this species takes the pressure off the rare endemic hardwoods.

Arabuko-Sokoke forest, Kenya
This is a protected coastal forest, one of the largest in the region. It is rich in rare species, but the trees are under threat from illegal logging for woodcarving. These forest refuges contain many plant and animals unique to the area, including the Sokoke scops owl (Otus ireneae) and the golden-rumped elephant shrew (Rhynchocyon chrysopygus), both endangered species threatened by loss of their forest habitat.

natural forest. Recycling of waste paper could be done on a far greater scale than at present; straw and sugar cane residues (bagasse) can be used to make paper too, and annual crops could be taken off esparto grass, reeds, bamboos, fibrous hibiscus species (already being used in the southern US), and residues from hemp fibre-making. 4,000 hectares devoted to raising hemp could have a pulp-producing capacity equal to 16,000 hectares of pulpwood forest.

Experiments with hybrid poplars have shown that eight crops can be taken from plantations of 9,000 trees per hectare. Alternatives to trees are, however, preferable because wood-pulp processing effluents cause much more serious pollution than other fibrous sources.

Wood is capable of being cut, turned, shaped, or drilled into almost any form. It provides the spears, harpoons, bow and arrows of the hunter; bowls, dishes, eating utensils, and boxes for domestic use; children's toys, cricket and baseball bats, hockey and polo sticks, tennis racquets, oars and paddles, grand pianos, violins, and nearly all our furniture.

All transport was originally wooden. The first vehicles were sledges, used as much on land as on snow. Wheels were first fitted to such sledges, and from these the two- and four-wheeled carts and wagons developed. Until the nineteenth century every kind of boat had been made of wood, too, from Neolithic dugout canoes to the huge warships and transports of the sixteenth and seventeenth centuries. In Britain, oak forests were planted and cut down to maintain the navy. The historic long sea journeys that

resulted in the early movement of plants around the world were all by wooden ship.

Early aircraft were based on wooden frameworks, and the Mosquito, using balsa, was one of the most successful fighters of World War II. Balsa (Ochroma) is the lightest important commercial wood, with a specific gravity of 0.12. It is a favourite material for model-makers as it is so easily cut. (The heaviest wood is probably South American ironwood, Krugiodendron ferreum, with a specific gravity of 1.3, which sinks in water.)

Wood is one of our most basic building materials, and the obvious choice wherever trees grow. In some traditional communities, branches form the support for coverings of hides, cloth, palm fronds, bark slabs or woven matting, from small shelters and teepees to the great communal houses of South America and New Guinea. Some New Guinea houses were based on huge mangrove trees fixed upside down so that their spreading roots formed the support for the roofing material.

As tools and craft developed, shaped timber formed the structural basis of most buildings, even those as big as the Cretan Palace of Knossos. In dwelling houses, the timber structure might support wooden planking for walls, daub and wattle (itself a wood product), or bricks and mortar. (Bricks, incidentally, traditionally included straw to improve strength; straw of teff, Eragrostis teff, still the most important crop in Ethiopia, went into making bricks for a pyramid over 5,000 years ago.) Wooden shingles, usually of cedar, went on the roof as tiles. Many present day houses are still wholly or partly of wood.

Mahogany is one of the finest timbers in the world, with a long and colourful history. The original mahogany was used by the people of Hispaniola to build canoes, and so was chosen by Spanish explorers to repair their sailing vessels in the West Indies. True mahogany (Swietenia spp., especially S. mahagoni) is ideal for everything from pianos to ships. There are several tropical trees known as mahogany , some of which are not even closely related. In the late seventeenth century a similar wood entered trade, originating this time from West Africa. It came from Khaya species which are close relatives of Swietenia, both in the family Meliaceae. Trade in both American and African mahogany continued to expand into the latter half of the twentieth century. In 1966 British traders imported 79,000 cubic metres at £38 a cubic metre, which rose to 196,000 cubic metres at £230 a cubic metre by 1979. Dwindling supply of these mahoganies had forced prices up. Tragically, true wild mahogany is restricted to fragmented populations, and remains under threat, although plantations now help to supply the trade. Already commerce is looking elsewhere, to ravage another set of species. The timber of dipterocarps (such as Shorea species), traded as Philippine mahogany, is currently being used for products at the cheaper end of the mahogany market. Some of the species involved were unknown in international trade before the 1950s. One wonders how long they will last. In Brazil, illegal exploitation, especially of mahogany, in the Indian lands has led to violent conflict between the loggers and native people, resulting in forest destruction and damage to traditional societies.

Flax (Linum usitatissimum), **above**, *was woven into linen in Babylonia and Egypt around 3000* BC. *This dual-purpose plant also yields linseed oil. The strong fibres, up to a metre long, have special qualities of water absorption, drying and gloss. Knowledge and use of plant fibres is still strong in many cultures; this Machiquenqa Indian woman of Peru,* **left**, *is weaving "cushma" to a traditional design.*

An important stem fibre is jute (Corchorus capsularis) *a cultivated plant about three metres tall, mainly grown in the Ganges delta of India. Too coarse for clothing, jute fibre goes into burlap or hessian, and is used for sacks and as backing for linoleum and other flooring. Jute only became established as a commercial crop in the mid-nineteenth century, after the Dundee spinners had learned how to use the fibre.*

Mahogany plantations have been established in several countries, including Fiji, Trinidad, Honduras and India, but a semi-mature plantation of *Swietenia* in Fiji was devastated by shoot-boring insects and had to be pulped. This is always a risk with monocultures, especially when clonal propagation of a high-quality original means that all the plants are genetically identical.

Fashions and fibres

Every tree and flowering plant contains fibres, which give it strength to stand and resilience to bend with the wind. People soon took advantage of this, twining sturdy ropes out of climbing stems, bunches of grass, and roots. The fact that their strength came from internal fibres was probably discovered when foliage or stems rotting in water were observed to leave a skeleton of tough threads. The idea of strengthening these fibres by twining them together to form yarn then led naturally to making this yarn into ropes and cords, interlacing strands to form baskets or, by weaving, into mats and cloth.

Weaving is surprisingly ancient: there is evidence of it from about 5000 BC. An Egyptian predynastic dish of about 4400 BC depicts a horizontal loom, the type that can be fixed to legs on the ground, or even between a tree or pole and the weaver's waist. You can still see such looms in Mexico today; they are ideal for nomadic people, who can simply roll up the weaving on its two end struts when they move on.

Although all plants contain them, only a few natural fibres have proved really useful to us. Wood fibres are so tough that they can only be separated by powerful machinery and are then too short for anything beyond the microscopic meshing which makes paper products. Flax, hemp, jute, cotton and sisal, on the other hand, have great importance.

Hemp *(Cannabis sativa)* appears to have been grown in China at least 4,500 years ago; several millennia later the original jeans were also made of it. However, the use of hemp for cloth declined because its cultivation was banned in many countries, as the species is also the source of the drug marijuana. But now hemp is undergoing a resurgence as a fibre crop, since new varieties have been bred with more fibre and a very low content of the active chemical THC. The main producers now are China, Romania, Spain, Chile, Ukraine, Hungary and France. In many ways, hemp is an ideal crop, requiring no fertilizers or pesticides, and growing fast. Hemp seeds are also used for flavouring, and have medicinal properties.

Manila hemp also produces fibre, partly for textiles, but mainly for making paper. This plant is actually a species of banana *(Musa textilis)* native to the Philippines. The fibres are extracted from the edges of the leaf-bases.

Another important leaf fibre is sisal, which comes from *Agave sisalana*, a plant of Mexican arid lands. Its long harsh leaves are cut from the base of the rosette over several years, until the plant flowers and dies. The leaves are crushed and cleansed of pulp by machine. Sisal fibres are a metre long, very strong, but generally too coarse for clothing; they go mainly into cord and binder twine.

Brazil produces the most sisal (over 180,000 tonnes), the other producers including China, Mexico, Kenya and Tanzania.

It is the versatile cotton plant (p. 20) which leads the field in clothing fibres. Very different from other plant fibres, cotton is produced on hairy seeds of species of *Gossypium*, relations of hibiscus and hollyhock. Each cotton fibre is a single cell, 3,000 times longer than thick, and a kilo of cotton contains roughly 200 million seed hairs up to five centimetres long. With cotton, the tedious processing of leaf and stem fibres is unnecessary: it has simply to be spun into thread.

A similar seed fibre is that of kapok or silk-cotton tree, *Ceiba pentandra*. Though not suitable for spinning, it is used to fill jackets and other cold-resistant clothes, mattresses, quilts, and sleeping bags; buoyant and impermeable to water, kapok is an essential ingredient of lifejackets.

In industrialized countries, synthetics have taken over from many natural fibres, often being cheaper, harder wearing, and able to resist insect and bacterial attack. They are not always so comfortable, however: the label "100 percent pure cotton" is still proudly carried by some (usually more expensive) garments. In many more basic cultures, natural fibres are still depended on for both cloth and cordage.

One remarkable plant still used in cloth manufacture has remained a step ahead of any technological equivalent – the fuller's teasel *(Dipsacus* spp.*)*, with its egg-shaped flower head, each floret surrounded by long, fine, spiny bracts. Mounted on frames or boards, these heads are still used to raise the nap on some cloths; for most fabrics wire substitutes will do, but for top-quality dressed cloths nothing surpasses the elegant teasel.

The renewable and the irreplaceable

Plants, in their infinite variety, provide wealth to large sectors of humanity. But we seldom distinguish between short-term and long-term gain. Domesticated crops, occupying land irrevocably transformed from its original plant cover, can be considered renewable assets – as long as these croplands are well managed. But fortunes today are also made from wild plants – and these are often irreplaceable. If we continue to exploit them at current rates, wild species and the vegetation they comprise – particularly tropical forests and savannahs – are, in fact, non-renewable resources.

But wealth from the wild *can* be renewable, with proper management. Rattans are a marvellous example: these multi-purpose tropical climbing palms can be sustainably culled from managed forest, and are the second most important item of trade from the rainforests of South-east Asia. The stems are used to make ropes, baskets, matting and furniture; the spiny leaves are made into fish traps and anti-theft devices. Once the spines are removed, the leaves are split and woven to make window blinds and thatching. Rattan production tends to be a village-based industry chiefly during slack agricultural periods, so local people can benefit from it more than from commercial timber operations. Since rattan extraction does almost no damage to the forest structure, the cottage industry can operate within forest reserve areas; and, since it provides an income, it may be a useful bargaining piece for conservation – for naturally, if the forests disappear, so do the rattans.

This type of sustainable forest exploitation can include a huge variety of commercially valuable substances – Indian forest plants, for instance, yield medicines, spices, foods and beverages, forage, gums, oils, resins, tanning and dyeing materials, pesticides, honey and other animal products.

In China, vast areas of subtropical hillsides are given over to the cultivation of managed forests of tall, multi-use bamboos, especially moso bamboo *(Phyllostachys pubescens)*. These tree-sized giant grasses are harvested on a sustainable rotation, for a huge range of useful products: poles for scaffolding and building, furniture, edible shoots, "Venetian" blinds from split canes, and even toothpicks.

Another type of wild vegetation offers great potential for renewable use – the ocean flora. Seaweeds provide ingredients for a huge range of products in industry – from fertilizer to food stabilizers, emulsifiers and gels; from microbiology to shampoo; from paints and dyes to deodorants, even building materials. Giant kelp is now being tested for alcohol production, offering a vast biomass resource. As for medical uses, research is only just beginning. Seaweed has been largely a wild harvest – but Chile, the Philippines and China now practise "mariculture" – they are farming the wild plants of the sea.

*Seaweeds manage to subsist on rocky tidal coastlines, perhaps the most changeable environments in the world; beyond the range of tides in some regions there are forests of kelp, as on the Falkland Islands, **right**. Individual plants may be over 50 metres long – the giant kelp (Macrocystis pyrifera) is reputed to reach 200 metres on occasion.*

Some species of bamboo are used to make a staggering range of items. In China, for instance, bamboo has been a basic raw material for rural societies for thousands of years. Its lightness, combined with strength, makes it ideal for constructing furniture, and even small huts and houses.

Little is wasted from a bamboo crop. The smaller canes can be split and woven into fences, or, as here, sewn together to make "Venetian" blinds. Even the ends of the canes are swept up from the workshop floor and used to make toothpicks.

This concept of a sustainably managed ocean harvest is a very far cry from the senseless gobbling up of every vestige of green growth in a forest by pulping machines, leaving nothing behind. In terms of wealth, it makes no sense at all to continue using up the irreplaceable. It makes very good sense to concentrate instead on exploiting all possible uses of plants we already cultivate. And for the wealth of the future, there is a deep necessity to resuscitate land which has been over-cropped, or ruinously denuded of its original cover.

The modern entrepreneur, intent on opening up the country for ranching, alcohol factories, and oil and mineral exploitation, says of the irreplaceable tropical jungle he is cutting down, "Here there is nothing." But conversely the native will say, "The forest has everything. It supplies all our needs, the food we eat, timber and thatch, wood for bows and arrows, fruit and medicine." The forest is either nothing or everything. It depends who you are.

Such knowledge was until recently common to people who lived in the forests, all too few of whom remain. A Peruvian tribe, for instance, has been recorded as familiar with the characters and uses of 50 species of palm tree alone. Before it is too late, people need very urgently to record this knowledge for posterity.

The greatest value of marine plants to humanity is as the foundation of food webs which support fish, the wild harvest of the oceans. But they have many direct uses too. Cast-up seaweed is applied as a manure in coastal regions, and seaweed products – alginic acid, agar, carrageenin and gelans – are widely employed in the food industry. In textiles, seaweed products are used to condition cloth; and they have numerous uses in medicine and cosmetics. Agar is probably best known as a microbiological culture medium for growing fungi and bacteria and for tissue culture. Kelp harvesting provides coastal peoples with an income in many areas, whether as a cottage industry or as a commercial venture, using modern cutter boats, **left**.

The chart shows the long history of medicinal plants as recorded. The oldest traditions come from China, Sumeria, Egypt and India. In Europe, the Dark Ages produced a long period of ignorance from which modern western medicine slowly emerged. Today eastern and western traditions are converging – producing a dynamic interaction of ideas out of which a new medicine may be born.

China *Around 3000–2730 BC the legendary Emperor Shen Nung composed the Pen T'sao Ching, with 365 herbal remedies. Ginseng was prescribed from at least this time, together with opium and ephedra.*

Near East *Sumerian ideograms of 2500 BC list several medicines based on plants. Circa 2200 BC tablets record 1,000 medicinal plants. Around 2500 BC the Assyrians listed 250 medicinal species.*

Hammurabi of Babylon, who ruled 1728 to 1686 BC, listed many healing plants.

Hammurabi receives scroll from a god

Egypt *Dating from 1550 BC, the Ebers Papyrus gives details of Egyptian medicine, with material going back 5 to 20 centuries before. It has 877 prescriptions and recipes. Most are medicinal.*

Egyptian papyrus with charm for women's diseases

India *Dating back more than 3,500 years, India's medicine has an ancient history. The knowledge was formalized in sacred poems or Vedas.*

Ayurveda, as this system is known, is still widely practised. It contains over 8,000 herbal remedies, using some 1,400 species, and has spread across South-east Asia.

From 1563, the Coloquios of Garcia da Orta (who was Portuguese) present first-hand knowledge of the medicines of India.

Greece and Rome
Born in 460 BC on Kos, Hippocrates was the most distinguished physician of antiquity. He travelled abroad researching and practising medicine, leaving a legacy of some 400 remedies.

Hippocrates the Greek physician, founder of medicine

In 332 BC the medical school at Alexandria was founded. It attracted the foremost Greek scholars. Scientists travelled with the armies of Alexander the Great from Greece to India. The learning of these conquered countries was set down in 700,000 or more books. All were destroyed by a mob of Christian fanatics in AD 391.

Circa 120 BC Crateuas was physician to King Mithridates of Pontus. Only fragments of his work remain.

Born in AD 23 in Verona, Pliny the Elder was a prolific Roman writer. His life's work was 37 books, eight of which dealt with medical botany.

In the first century AD, Dioscorides wrote De Materia Medica. This described some 500 plants and was the standard source book for 1,500 years. He was an army surgeon.

Born around AD 131 at Pergamon, Galen travelled extensively, studying medicine at Alexandria. The allopathic and homoeopathic systems of medicine are based on his doctrines. His books were much copied during the Middle Ages.

A little known Roman of the fourth or fifth century, Apuleuis wrote a Herbarium of which Saxon and later translations exist.

Arabia *Beheaded in AD 303, St Cosmos and St Damian are the patron saints of surgery. They were Arab physicians who travelled extensively preaching Christianity and healing. Plant-based medicines were important in Arab culture from the earliest times.*

Born near Bokhara in AD 980, Avicenna was one of the most famous Arab doctors. His greatest work, the Canon of Medicine, refers frequently to Galen and Aristotle.

Arab physicians continued the tradition of medicine after the decline of Rome, translating the Greek books and adding their own observations. The culture ended with the Mongol invasion which took place in the thirteenth century.

From an Arabic copy of Dioscorides circa AD 1222

Europe *Written between AD 900 and 950, the Saxon Leech Book of Bald (a friend of King Arthur) was the first English herbal written in the native language.*

Founded during the ninth century AD, the first organized medical school in Europe was at Salerno. It gradually declined in the twelfth century.

During the fourteenth and fifteenth centuries, Montpellier had the foremost medical school in Europe. Others arose at Bologna, Padua and Paris.

Paracelsus (1490–1541), a Swiss alchemist, developed the *Doctrine of Signatures.*

Born in 1501 in Siena, Mattioli tried to reconcile the teachings of Dioscorides with the innovations of the Arabs and the new herbal remedies that were flooding in from across the world.

In his New Herball of 1551–1568, William Turner was the first Englishman to study plants scientifically.

In 1585 the Herbario Nuovo of Castore Durante was first published in Rome. It gave all the known medical knowledge of plants of Europe and the East and West Indies.

The first large-scale British herbal is that of Gerard, published in 1597 in two large, heavily illustrated volumes. It was largely based on the work of the Dutchman Dodoens.

More popular than Gerard was Culpeper, whose herbal of 1652, first entitled The English Physician, was condemned by contemporary medicos for his heavy reliance on astrological botany.

"Le Propriétaire des Choses", Paris 1500

South America *In the fourteenth century AD, the well-developed medical knowledge of the Aztecs was set down in pictograms. Most were destroyed by Spanish invaders.*

South American figurine chewing coca

Born circa 1514, Francisco Hernandes was dubbed First Physician of the Indies by Philip II of Spain. He wrote Rerum Medicarum Novae Hispanae Thesaurus *concerning Aztec medicine.*

In 1552 the Badianus Manuscript was compiled by two Mexican physicians at Santa Cruz describing and illustrating healing plants used by Aztecs.

In Chronica del Peru *of 1553, Pedro de Cieza de Leon mentions çoca among other plants.*

First printed in 1569, the Dos Libros *of Nicholas Monardes, are concerned with the medicinal materials of the West Indies and New Spain as reported to Monardes in Spain.*

North America *Seventeenth-century settlers imported European plants. They used* Culpeper's English Herbal *extensively. They also experimented with herbs well known to the native Indians, some of which they exported back to Europe.*

In 1718 the Canadian Jesuits started to export native ginseng to the Orient to supplement the scarce Chinese supply.

Native North American medicine man

Australia *Written herbal history dates from the first British settlements in the late eighteenth century. Only a few Aboriginal herbs have entered mainstream herbalism.*

Africa *There are no written records so one can only surmise the antiquity of African herbal learning. The oral tradition is still extensive, but its value is only now appreciated.*

CHAPTER SEVEN

Green Medicine

Green medicine is very old. Five thousand years ago, the civilizations of antiquity knew plants such as *Ephedra* and the opium poppy, still used for the drugs ephedrine, morphine and codeine. It is also very young: scientific investigation into the medical properties of plants, from cancer cures to contraceptives, is today sparking off a revolution in our understanding. Interest in eastern drugs such as ginseng, and in the disappearing plant lore of indigenous cultures, is matched by concern for the unexplored potential of tropical species now threatened by forest clearance. The gap between western and traditional medicine is closing at last.

Only recently some scientists believed that medicinal plants were a thing of the past. Since some drugs originally derived from plants can be synthesized directly far more cheaply and easily, it was felt that once the secrets of their chemical composition had been discovered, medicinal plants would become redundant. But in the early 1970s attitudes started to change. On the one hand many people in developed countries, especially North America and Britain, became curious about herbal medicine and welcomed it as part of a "back to nature" philosophy as well as for its lack of side effects – a disadvantage of many powerful pharmaceuticals. And in many developing countries, policy-makers started to encourage a revival of traditional medicine using local herbs, because it was cheap, available and a part of their culture. Furthermore it reduced dependence on imported drugs. Science is now looking at herbal medicines in a more systematic way, and screening many plants for active substances.

Meadow saffron (Colchicum autumnale) *is the main source of colchicine. This poisonous painkiller is remarkably specific: excess causes diarrhoea or even death but in small doses it relieves the pain of gout. Unfortunately though, it does not reduce the high level of uric acid which causes the affliction. Various species of* Colchicum *have been used for over 2,500 years. The resemblance of the root of meadow saffron to a gouty foot earned it a long folk reputation.*

Magic to medicine

Early peoples must have shown an insatiable curiosity in sampling and testing plants, for they acquired over the millennia of prehistory an intimate knowledge of many thousands of species. Besides food plants, there were those that could dull pain, heal wounds, cure fever, or give energy and strength. Others could create hallucinations, induce sleep, or calm the spirit. And some were potent weapons – including poisons which killed instantly. Since certain plants could kill, create visions, or cure depending on the amount taken or the part used, their powers were seen as magical or divine: thus magic and medicine became inextricably linked, as the preserve of "witch doctors" and shamans. Immensely powerful in their communities, these knowledgeable specialists were more often interested in the spiritual cause of illness than in diagnosis, so the nature of plant medicines remained obscured by ritual and superstition.

Over the centuries, knowledge grew and was passed on orally from generation to generation; in Africa and South America for example the oral tradition is still strong; in Asia and eastern Europe written records – the great "pharmacopoeia" – began very early. In western Europe, plant medicine remained as folklore until the birth of organic chemistry in the nineteenth century. Only then did the curative properties of plants, the compounds involved and their modes of action begin at last to be unravelled, as the scientific tradition emerged.

A curious example of how modern knowledge is born of apparent superstition is given by the *Doctrine of Signatures*. Widely invoked by societies from the ancient Chinese and North American Indians to the Europe of the Middle Ages, it taught that like cures like: a heart-shaped leaf must be good for heart disorders, kidney-shaped for kidneys, a yellow juice for the treatment of jaundice. Most of these "cures" are unsubstantiated; but some plants have been found to be useful in other ways, while one or two actually deserve their reputation. The trembling leaves of white willow resemble the shaking of a fever patient; a derivation from its bark, long since used in the treatment of fevers, now has its place in the blueprint for aspirin. And the meadow saffron, with a foot-like projection on its bulbous root, proves to be effective against the pain of gout.

Modern medicine, with its thriving pharmaceuticals industry and bright technology, tends to make us forget that the average medicine off the shelf was probably first made, not by a chemist, but by a green plant. The current *British Pharmacopoeia* contains no fewer than 80 plant genera – about 30 percent of modern prepacked medicines are based on plant products while 80 percent or more were plant-based at some time. But this is a small percentage compared with the range of healing plants known to traditional medicine.

The healing potential

Plants are extraordinary chemical factories. Their storage of sunlight energy, their daily growth, their flowering and seeding, are complex enough, involving the syntheses of a huge range of organic molecules. These primary activities are common to all plants. However, individual plant species produce many other specialized substances, called secondary compounds, which differ from plant to plant depending on their circumstances. Plants do not flourish in isolation, but in intimate relationships with many species, from those that feed off them or attack them to those that help their pollination or seed dispersal. Some of these chemicals are therefore defensive, some advertise the plant's presence or play a role in symbiosis, some are waste products or toxin neutralizers. Many are of great value to man – for instance as hormones and antibiotics, narcotics and antihistamines, pesticides and poisons.

Certain plant families are strong in one or more types of secondary compound. Within a group of closely related species, some small changes in genetic instructions will cause slight differences in the chemistry so that the chemicals produced may have

White willow
Salix alba

female

male

Meadowsweet
Filipendula (Spiraea) ulmaria

dramatically different effects. Such changes are of immense interest to the pharmacist.

Three very important groups of secondary compounds are the alkaloids, glycosides and saponins. Alkaloids, nitrogenous compounds such as caffeine, morphine, quinine and strychnine, many of which affect the central nervous system, can be lethal if large amounts are taken. Deadly nightshade and the opium poppy are alkaloid-bearing plants.

Glycosides are a type of secondary compound in which a physiologically active molecule is linked to a sugar molecule. Such compounds include the cardiac glycosides valuable in heart treatment extracted from foxglove; purges like cascara, senna and rhubarb; and the saponins, which foam or froth in water like soap – soapwort can literally be used to wash the hands.

Plant compounds are big business: in the United States, herbal medicine use rose from 2.5 percent of the population in 1990 to 37 percent in 2000, and has risen further since. Every year the US imports hundreds of thousands of tonnes of herbs to support its $3 billion dollar market, and Germany, France, Italy, Spain and the UK are also major importers. Europe imports a quarter of the

world's trade in herbal medicines – about 440,000 tonnes. Britain spent £65 million on herbal remedies in the year 2000, and this market has been growing at up to 20 percent each year.

Some plant compounds are isolated by laboratory techniques and made up into medicines; some are completely synthesized in the laboratory. But we still have much to learn. Some compounds – the cardiac glycoside from foxglove is one – have so far resisted attempts at synthesis and it is still necessary to use plant material as the basis of every dose administered; others, as in the case of curare (see p. 114) are more effective than their synthetic derivatives. The drug in aspirin, on the other hand, probably the most widely taken medicine in the world, is now entirely synthetic. Plants have created a quarter of all prescribed medicines and 100 species form the basis of modern drugs.

In developing countries, plants are usually the main source of medicines, and up to 80 percent of the world's people rely on traditional, mostly plant-based, medicines for their primary health care. Over 35,000 species of plant are used for medicine, including at least 7,500 from India alone. Not surprisingly, the increasing use of

The strange story of aspirin is typical of plant research and discovery. Since Dioscorides, it was known that the bark of the White willow (Salix alba, **above left**) *could kill pain. Decoctions or infusions of it were used against fevers, rheumatic pains, gout, toothache, earache and headache. The active ingredient was first isolated from willows in the nineteenth century and named salicin. A similar compound, salicylic acid, was isolated from Meadowsweet,* Filipendula ulmaria *(then called* Spiraea ulmaria, **above right***). In 1899 it was found that the combination of salicylic acid and acetic acid – the chemical acetylsalicylic acid – was more effective. This substance was then, confusingly, named "aspirin" after* Spiraea. *Salicylic acid is still in use today against athlete's foot and skin diseases such as acne.*

The mandrake's remarkable career started with the resemblance of its forked tuberous roots to a human body. Barren women in its native Near East hung the roots up in their houses. Rachel's barrenness was ended with the birth of Joseph after she had begged her sister "Give me, I pray thee, of thy son's mandrakes" (Genesis 30). From preventer of barrenness to aphrodisiac: the Romans reputedly used mandrake in orgies.

The magic that came to surround mandrake (Mandragora officinarum) *is extraordinary: in the first century Josephus wrote, "He who would take up a plant therefore, must tie a dog thereunto, to pull it up, otherwise if a man should do it, he should surely die in a short space after."*

The narcotic properties of mandrake were more real: Dioscorides wrote that it was given "to such as shall be cut, or cauterized … For they do not apprehend the pain, because they are overborn with dead sleep." Greek, Roman and Arabian doctors used it as anaesthetic and to calm maniacs. In Palestine when the Romans started to crucify criminals, relations offered a sponge soaked in mandrake juice and placed on a pole for the men to drink, so that they became unconscious and were mistaken for dead.

Mandrake was long an important herbal plant but its basic alkaloid, mandragorine, is not now used medicinally.

wild plants for medicine has led to conservation problems in many cases (see Herbal medicines and conservation, p. 128, for more on this topic).

Poisons and panaceas

However they are prepared, our powerful modern plant-based drugs have usually reached our notice by one of two routes: either their reputation as healers in folk medicine provoked scientific investigation; or they were originally known as dangerous poisons, and then tested for remedial possibilities because plants with the capacity to kill can often cure. Knowledge that a plant contains strong toxins is today a spur to investigation, working on the principle that what can destroy or paralyse in large quantities may also have a useful action in small concentrations.

A striking example is the calabar bean *(Physostigma venenosum)*, once used for the ritualistic practice of trial by ordeal, in which the accused had to imbibe a poisonous liquid; death meant guilt. (Those who knew the ropes gulped the liquid down and promptly vomited; the fearful sipped it slowly and died quickly of asphyxiation from paralysis of the respiratory muscles.) Synthetics derived from its alkaloid, physostigmine, are now used to build up muscle strength in the disease myasthenia gravis and, especially, to treat eye disorders. In these respects it is similar to an even more valuable eye medicine, pilocarpine, derived from the jaborandi tree *(Pilocarpus jaborandi* and spp.*)*, considered by the World Health Organization to be one of the most important plant-based drugs in the treatment of glaucoma.

Toxins used for hunting or warfare have also yielded some powerful medicines. South American Indians made one such poison, curare, from several plants including species of *Strychnos* and *Chondrodendron*, and used it both for war and to kill game. Not until 1939 was the active principle of curare isolated – a muscle relaxant now widely used in abdominal surgery to facilitate cutting by relaxing striated muscle. The early Europeans tended to kill their enemies with poisons from the great Solanaceae family, with its roll-call of sinister names – mandrake, deadly nightshade and henbane. Every part of the last two is poisonous; they have a long record of use by professional poisoners. The chief symptoms were flushing, a dry mouth, dilated pupils, and delirium with death from respiratory failure. Interestingly enough, this family also includes the potato and tomato, both thought to be wholly poisonous when they were first brought to Europe from the Americas.

Today, the Solanaceae rank high in the list of potent healers. Extremely valuable alkaloids are found in deadly nightshade, henbane, thornapple and other *Datura* species, and the Australian *Duboisia*. Medically these are active against a large number of complaints, being useful in asthma, and for treating diseases of the central nervous system such as Parkinson's disease. Atropine and hyoscine are used in ophthalmology to dilate the pupil, and against stomach and duodenal ulcers. Hyoscine allays colic and other internal gripings, and is used in mania and alcoholic

Kill or cure

Hyoscyamus niger

Atropa belladonna

Datura stramonium

Deadly nightshade is named Atropa belladonna, *Atropos being one of the Fates who severed the thread of life, and belladonna meaning "beautiful lady" – women used to place a little dilute juice in their eyes to expand the pupil and give themselves a bright, wide-eyed look. This capacity is today valued by ophthalmologists, who need to dilate the pupil for examination. Deadly nightshade is especially dangerous because its bell-shaped flowers are followed by glistening black berries, sweet to taste, and tempting to children – to whom two or three are lethal. The Roman poisoner Locusta is reputed to have killed the Emperor Claudius with belladonna. Its effects start with hallucinations and delirium, then paralysis before death.*

Henbane (Hyoscyamus niger), **above**, *is a tall, thoroughly evil-looking, evil-smelling plant with large open bell flowers – veined with purple on lurid yellow in the north European species. Dioscorides recommended the milder white-flowered henbane of the Mediterranean (H. albus) as an "easer of pain" with many uses. In Europe later on, criminals used it to render their victims unconscious.*

Thornapple (Datura stramonium), **above right**, *an American native now widespread in many countries, has an attractive white trumpet flower succeeded by large, thorny seed capsules,* **right**. *Its American name, Jimsonweed, comes from Jamestown where sailors once nearly killed themselves eating what they believed to be spinach. Indian prostitutes doped clients with it then robbed them, and it also played a part in the white slave traffic.*

Opium poppies

disagreeable sensations apart from skin irritation. It is also a mental stimulant and addictive. Those addicts who try to give it up suffer appalling symptoms, both physical and mental, sometimes leading to death.

Codeine is a household painkiller used in cough medicines, cold remedies and for a host of diseases; about one-fifth as strong a painkiller as morphine, it is very much less addictive. Another important alkaloid of opium is the non-addictive moscapine, which relieves coughs, especially in bronchitis and whooping cough.

The two-edged sword of healing and destruction common to all these poisonous medicinal plants is wielded most strongly by the opium poppy (Papaver somniferum), **above**. It is the source of the painkiller morphine, which is difficult to synthesize commercially. Heroin is made from morphine.

In its crude state opium has been known for several millennia in the west for its sleep-inducing and painkilling qualities. Among the medicines made from it were paregoric, given to calm babies and make them sleep (still available) and laudanum (now called tincture of opium), widely prescribed as a painkiller in the nineteenth century. Coleridge wrote Kubla Khan under the influence of laudanum; Elizabeth Barrett Browning became an addict while under medical attention; and in the twentieth century Cocteau's hallucinatory drawings and writings reflect its influence.

Opium is a mixture of some 25 alkaloids, including morphine. Named after Morpheus the Greek god of dreams, this is the greatest natural painkiller of all and its medicinal value in extreme cases has been considerable. Unfortunately its side effects can be most harmful and for this reason its use is now restricted. It relieves pain and resulting sleeplessness, calms anxiety, and reduces all

Native to south-eastern Europe and western Asia, the opium poppy is now widely cultivated from Turkey eastwards, **above**, especially in Afghanistan, India, Myanmar and Thailand. Some governments are trying to encourage the cultivation of alternative crops to supplant their peoples' economic dependence on harvesting this deadly crop.

The Greeks celebrated the opium poppy and its calming powers in literature, art and religion from earliest times. The Cretan Poppy Goddess, **left**, dates from the later Minoan period, 1700–1400 BC. The Greek culture regarded sleep as the greatest healer, hence the frequency with which poppy heads are found on figurines, vases, utensils, jewellery and coins. The Greeks also called opium "nepenthe", and in Homer's Odyssey Helen offers Telemachus a philtre of this to "lull pain and bring forgetfulness of sorrow".

delirium and as pre-anaesthetic medication. Its most familiar use is for travel sickness.

To produce these important drugs, night-shade and henbane are now grown commercially in North America and the Balkans, as are species of *Duboisia* in Australia. A good deal of atropine still comes from the Egyptian henbane *Hyoscyamus muticus* in the wild. Although atropine can be synthesized, all the drugs are still made from the starting point of the dried leaf, flowering tops and roots of the various plants.

Many other plants as poisonous as these were once harnessed as medicines, but are now considered too dangerous to use – for instance the buttercup family, which includes such popular garden plants as helle-bore, delphinium, columbine and monks-hood or "wolfsbane" – a name in keeping with its lethal character. Ironically the peony, named after Paeon, a favourite physician of the Greek gods, has also dropped out of use in the west despite a long tradition; but it remains in Chinese herbalism.

Quinine – the "bark of barks"

After painkillers, the plant cures which have probably done most to relieve human suffering are the febrifuges (anti-fever), and the greatest of these is quinine, which acts against malaria.

This widespread fever – "the ague" – has had an extraordinarily harmful effect in human history. It regularly ravaged Rome, which Cicero called "a pestilential city"; the fall of Rome and of ancient Greece have both been attributed to malaria. Alexander the Great died from it. Africa repelled European settlers for centuries because of its prevalence. Ague also once affected people in the fenland of south-east England. Many fear that with global warming, the disease could spread north again. Until quinine became widely available, malaria probably killed two million people a year, half of them in India, and left perhaps 200 million chronically ill. Not until 1902 was it known that the organism causing malaria was spread by mosquitoes. And not until after World War II did DDT enable us to tackle the mosquitoes directly.

Quinine originated from the South American Andes, from the bark of *Cinchona* trees, which was first brought to Europe in the seventeenth century. In 1820, the alkaloid in the bark was isolated, and demand increased vastly with foreseeable consequences – the trees with the bark containing the largest quinine content, in parts of Bolivia and Peru, were steadily destroyed by the Indian collectors. Both the British and Dutch, with malaria rife in India and Java, tried to collect seeds, young plants and root cuttings and grow them on in those countries. Unfortunately the resulting trees – over a million in the case of the British expedition – were worthless. Clearly, very few *Cinchona* plants contained worthwhile quantities of quinine. Eventually an Englishman obtained seed (illegally) from a grove of trees in Bolivia he knew were very high-yielding, and the Dutch acquired a pound of them for a paltry sum. From this single source, 12,000 trees were raised which yielded quinine in far higher proportion than had ever been recorded and

Quinine, still the most effective anti-malarial agent, comes from the Cinchona *tree of the Andes,* **above**. *The Jesuits of Lima first introduced the ground bark to Europe. Known to the Indians as quinaquina, "the bark of barks", quinine is obtained by stripping the bark from the trees, shown* **below** *– a process which eventually kills them. Some other alkaloids, such as quinidine also have pharmaceutical applications.*

formed the basis of a multi-million guilder monopoly – until the commercial production of DDT and synthetic anti-malarial drugs took over. However, by 1984 the malaria parasite had become resistant to most synthetic drugs in many countries, and quinine is returning as an effective remedy, while other plant remedies are also being investigated. About 90 different species are used to treat malaria by local people in Roraima, Brazil, but commercial drugs have yet to be prepared from them. A wormwood, *Artemisia annua*, is used traditionally in China and has proved to be very effective against some types of malaria.

Multiple saviours

Some plants have long stood out in the pharmacopoeia as multiple healers, effective against many ills, and a number of these have yielded modern drugs with correspondingly diverse uses.

One of the earliest was *Ephedra (ma huang),* used by the Chinese 5,000 years ago; they made a tea of the twigs to allay fever and coughs, and to increase blood pressure. Today the intrinsic alkaloid, ephedrine, is a valuable multi-purpose drug, still made from the plants (which are cultivated) but also synthesized. Ephedrine helps those with low blood pressure, and also has a stimulant action used to treat patients in comas. It prevents the fall in blood pressure which occurs after spinal anaesthesia in surgery, and contains the dizziness and fainting which are symptoms of "heart block". But perhaps the most generally appreciated virtue of ephedrine is in treating hay fever and asthma, for it clears both the nasal and the bronchial passages.

By contrast to ephedrine, reserpine – another long-known plant drug – lowers blood pressure, controls arrhythmia and calms the central nervous system. Reserpine comes from the roots of snakeroot *(Rauvolfia serpentina).* The powdered root has been in use for at least 2,000 years in India for those suffering from "moon disease" (lunacy) or other mental illnesses. Not till 1952, however, when the alkaloid was isolated, did its use in western medicine begin to snowball. Probably one of the most effective plant-based tranquillizers, it made the lives of schizophrenics far more bearable. Nowadays, the drug is mainly used to control high blood pressure. Snakeroot is collected on a large scale in the wild, mainly in India, Thailand, Bangladesh and Sri Lanka, partly for local traditional medicine, and partly for export. Unless this trade is regulated, it could become threatened.

Scores of local folk names attest to the foxglove's long-standing reputation as a multiple healer; among its many imagined medical virtues were efficacy against fevers and colds, treatment of the King's Evil (scrofula) and, notably, cure of dropsy. It was an eighteenth-century doctor, William Withering, who first suspected that foxglove acted on the heart, but only after his death was it realized that heart disease caused one kind of dropsy.

Today the foxglove's reputation is justified by results: probably millions of heart-sufferers owe their lives to this charming wild flower. Its active principles, the glycosides digoxin and digitoxin, can strengthen and regularize the heartbeat. They are still derived from the dried leaves, sometimes from *Digitalis purpurea* but now, mostly, from *D. lanata*, both native to Europe; the drug has never been synthesized.

Breakthrough on cancer

Many people know the rosy periwinkle (*Catharanthus roseus*) as a small pot plant, with showy pink or white flowers. Originally from the tropical forests of Madagascar, this little member of the poisonous and drug-rich dogbane family, has a long folk history against diabetes: a tea is infused from the leaves, and patent medicines containing it are still used in Madagascar and the West Indies as anti-diabetics.

Not surprisingly this reputation became the target for scientific testing. In the 1950s a medical research laboratory in Ontario, searching for a substitute for insulin, tested the plant's effect upon blood sugar levels in rabbits. First they were fed with the leaves; then they were injected with an extract from the plant sap. The tests were negative, but many experimental animals died mysteriously. They appeared to have lost resistance to a common bacterium – the periwinkle extract was reducing their white blood cell count and bone marrow activity. Quite independently, a botanical drug-screening company subjected leukemia-infected mice to rosy periwinkle extract – and found that their lives were significantly prolonged. The research teams pooled their knowledge, and in 1958 the substance active against the leukemia cells was isolated – an alkaloid, now called vinblastine. In the 25 years since,

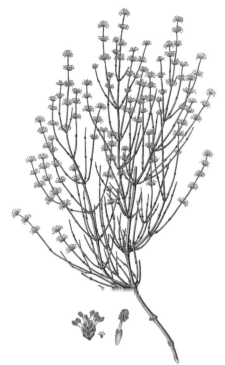

The value of ephedrine, an alkaloid extracted from species of Ephedra, **above,** was discovered for the west by two American pharmacologists who visited a Peking pharmacy in 1923. Their research into Chinese medical literature, and experiments, confirmed the efficacy of the Chinese Ephedra extract, ma huang. It is used mainly to treat asthma, hay fever and colds.

Rauwolfia serpentina, **below,** grows from India to Indonesia; the anti-hypertensive reserpine is obtained from its roots.

The foxglove (Digitalis purpurea), **above,** has a long history of herbal use, and it was this that led to discovery of its potential as a heart medicine. William Withering, who spent much of his time treating the poor, visited an old woman suffering from dropsy, seemingly far gone. A few weeks later, he found her much improved; he discovered she had taken a herbal remedy and that one of the herbs in it was foxglove. Withering spent ten years looking into its potential against dropsy, before he suspected its real value. Much later it was confirmed that foxglove acts on the heart; the improved circulation that results steps up the performance of the kidneys, which in turn clear the accumulated body fluids which our ancestors called dropsy. Foxglove was not apparently known to the ancients. Instead they recommended for heart problems monkshood (Aconitum), false helleborine (Epipactis helleborine) and the sea squill (Scilla (Urginea) maritima) of Mediterranean coastal areas – all highly toxic plants. Foxglove is poisonous too: Withering spent a long time getting the dosage roughly right. An excess leads to a number of unpleasant symptoms.

over 70 alkaloids have been isolated – no easy matter since, unlike other alkaloid-containing plants, the rosy periwinkle's total alkaloid content is minute – around 0.00025 percent of dry weight. Fortunately the drug is potent in tiny quantities. Vinblastine was first used in 1960: the patient's large cancerous tumour rapidly disappeared but, after two and a half years' treatment, cancer symptoms began to resurface. Then another periwinkle alkaloid, vincristine, was tried. Remarkably enough, it started again to control the disease, showing that built-up resistance to vinblastine was not carried over to vincristine.

These remain the two major cancer treatments derived from rosy periwinkle, and the plant is now grown commercially for medicine and well as for horticulture. Vinblastine – used in conjunction with other drugs – now produces a remission rate of 80 percent in Hodgkin's disease, compared with 19 percent from previous treatment; vincristine provides over 90 percent remission in acute lymphocytic leukemia, which frequently attacks children. Other types of cancer can also be treated, and further periwinkle alkaloids are being tried. All share the property of interfering with cancer cell growth.

The hunt for new drugs to fight this disease has involved screening many thousands of plants. The obvious place to begin was with traditional cancer cures in indigenous cultures: Penobscot Indians used the American mandrake *Podophyllum peltatum* against malignant tumours, other groups the bloodroot, *Sanguinaria canadensis*. These plants are now being exploited, both for drugs and for knowledge: some of the

synthetic analogues of podophyllotoxin are more effective than the natural compound. In India, the Himalayan *Podophyllum hexandrum* is the species used. Several other plants also show promise, including the East African shrub *Maytenus gratius*, the tall American meadow rue *Thalictrum dasycarpum*, and the Chinese plum-yew, *Cephalotaxus fortuni*. Anti-cancer drugs have also been isolated from species of yew (notably *Taxus brevifolia* and *T. baccata*), and also from hazel *(Corylus avellana)*. Happy tree (*xi shu*), *Camptotheca acuminata,* contains an anti-cancer chemical, camptothecin, which has been modified to produce many others, all being assessed and investigated for use in chemotherapy.

Destruction of vegetation makes the task of screening plants for useful chemicals all the more urgent. Recent studies sponsored by WWF have recorded over 1,000 plants in the rainforests of South America used by tribal people. Among them is *Croton lechleri*, whose bark is used to treat stomach cancer; another is *Jacaranda caucana* which is used for leukaemia. Loss of habitat threatens these and many other plants whose medicinal properties we may never know.

Contraceptive revolution

One of the most dramatic stories of the last half-century is the development of the contraceptive pill, which has revolutionized birth control. The American who made this possible, Russell E. Marker, was an organic chemist specializing in steroid compounds and, later, in the human sex hormones. Marker was looking for a drug to make up

The rosy periwinkle (Catharanthus roseus) is a relation of the hardy evergreen creeping periwinkles liked by Elizabethan gardeners, but has been separated botanically from their genus (Vinca) into the distinct tropical genus Catharanthus. Although naturalized in the tropics, it is often called Madagascar periwinkle from one of its truly native localities. However, little of the natural vegetation of Madagascar remains and most of the rest is in danger. A close relative of rosy periwinkle, Catharanthus coriaceus (also from Madagascar) is critically endangered from bush fires; it too may have valuable chemicals to offer – if it does not become extinct first.

The dainty little bloodroot (Sanguinaria canadensis), **right**, *so called because of the red juice that flows from the cut roots, was used by Native North Americans to treat several ailments. Modern science has confirmed its value against malignant tumours.*

for lack of natural progesterone hormone in women – a cause of miscarriages.

He began investigating plants producing steroid-bearing saponins, including the Joshua tree *(Yucca brevifolia)* of the American deserts and sarsaparilla *(Smilax* spp.) which had long been touted as a tonic and cure-all. Eventually he took the saponin now known as diosgenin, which the Japanese had isolated in 1937, from the yam *Dioscorea tokoro*, and showed how the female hormone progesterone could be synthesized from it in five steps, and the male hormone testosterone in eight. Lack of interest from major pharmaceutical companies provoked him, in 1942, into looking for yams nearer home in Mexico. Out of some 60 or 70 species he finally found four or five which produced tubers of reasonable size fairly quickly. These species provided the basis of the world's contraceptive pill industry – they were both extracted from the wild and cultivated on a large scale in Mexico. Today most oral contraceptive pills in the west are totally synthetic; but in India and China wild yams are still being processed for this purpose. *Dioscorea nipponica* and *D. zinziberensis* also contain diosgenin and are cultivated, especially in China which is now the biggest supplier of diosgenin.

There is a tragic footnote to this story. In northern India, on the foothills of the Himalayas, grows *Dioscorea deltoidea*, reputedly the best-known source of diosgenin: not only does it contain high amounts of diosgenin, but it is also free from minor steroids. Industrial companies have uprooted large amounts of its tubers from the forests,

ignoring strict regulations on exploitation; parts of Kashmir and Himachal Pradesh have been completely denuded. Even more serious, in areas whose plants remain, their diosgenin content is far lower than before, as big, high-yielding tubers are the first to go. One study found the diosgenin content reduced from over six percent to less than one percent. An answer, of course, is to domesticate the plant, breeding high-yielding varieties suitable for cultivation, and to conserve the full diversity of the wild ones in strictly managed reserves.

Marker also searched unsuccessfully for a plant source for the hormone cortisone. By then a well-known drug with enormous remedial value for rheumatoid arthritis, cortisone was being produced commercially from animal adrenal glands, but the process was complex and costly. No commercial plant source of cortisone has yet been found (an African *Strophanthus liana* does contain it, but not in sufficient quantities). However, anti-fertility drugs based on yams are helping women arthritics who take doses to control pain. This link between sex hormones and cortisone was discovered when women found their arthritis improved during pregnancy.

Many other plants have contraceptive potential, from the soybean, cotton, and fenugreek, to cures already known to indigenous cultures. In the species-rich forests of tropical South America, several kinds of tree bark, two arum relations *Dieffenbachia seguine* and *Anthurium kessmannii*, and numerous other herbal sources are used as contraceptives. In Colombia, a woman anthropologist who won the confidence of

American mandrake (Podophyllum peltatum), *also known as the North American mayapple, has long been used in the treatment of cancerous tumours and the resin is still the preferred therapy for venereal warts.*

The South African yam, Dioscorea elephantipes, **left**, *has potential for commercial diosgenin extraction and synthesis of sex hormones. It was in Mexican yams that diosgenin was first found in sufficient quantities to make the contraceptive pill available on any scale. Before Russell E. Marker took his two kilos of yam-derived progesterone to a large Mexican pharmaceutical company in 1942, the drug had been expensive and rare. Marker's yams at once broke the existing monopoly on commercial progesterone, and gave birth to the contraceptive pill industry. Among the many plants now being explored for hormone potential is the ordinary cotton plant* (Gossypium barbadense, **far left**). *Gossypol, extracted from the seeds, is being tested for use as a male contraceptive. Though long tried in China, however, clinical tests have yet to demonstrate that there is no risk to subsequent progeny.*

the Witoto tribe was shown a plant which was given to girls just once at puberty and prevented pregnancy for six to eight years. Another plant was alleged to confer permanent female sterility.

Oil from cotton seeds is a source of a male contraceptive, gossypol, and efforts are being made to produce such a pill commercially, although research showed problems, including permanent infertility. Another species may have potential here – the Chinese medicinal plant *Tripterygium wilfordii*, a species which also has anti-tumour and anti-HIV activity.

Plants against HIV/AIDS

HIV/AIDS is one of the most frightening diseases of the modern age, and it now affects over 40 million people worldwide, killing three million people in 2003 alone, and over 20 million since first identified in 1981. Huge amounts of money and effort are being put into researching cures and treatment. Nevertheless, the number of people with HIV continues to rise.

Plants are playing a role in this battle too – as the source of the rubber used to make condoms, which prevent infection from sexually transmitted diseases including HIV, but also as a source of potential anti-HIV drugs. One of these is the Moreton Bay chestnut *(Castanospermum australe)* which contains alkaloids that have been shown to attack the AIDS virus, reducing its virulence. Derivatives created from this compound are even more effective. Several other plants also have anti-HIV properties, including cotton (gossypol), poppies (papaverine), *Hypericum* spp. (hypericin), and the Samoan mamala tree *Homalanthus nutans* (prostatin).

Body as entity

There are two schools of thought on the role of medicine: one views it as curative, treating specific ailments; the other as preventive – keeping the body well-tuned and resistant to disease. From ancient times Chinese medical practice has been based on the preventive principle. Health-sustaining substances, mild and harmless

even in large doses, are regarded as "kings" among medicines, while curative drugs are merely "servants".

One of the main plant weapons available to the Chinese has been, for nearly 5,000 years, an ivy relation, *Panax ginseng*; its Latin name means "panacea", all-healing. This thick-rooted herbaceous plant of the ivy family surfaced in the west as an invigorating drug, reputedly preventing fatigue and depression, staving off the weakness of old age, increasing sexual potency; also improving headache, amnesia, tuberculosis, diabetes, heart problems and kidney disease, among others. Much empirical work done on ginseng bears out these claims. Ginseng and American ginseng *(Panax quinquefolius)* are both threatened by over-collection in the wild, although they are also cultivated for processing and for export. The Russian Siberian ginseng *(Eleutherococcus senticosus)* is closely related and is also the source of a tonic. Tradition has it that wild plants are more effective than those cultivated; the small gnarled roots indicate the plant has

Exports to the Orient, where sadly Panax ginseng *is an endangered species, are now threatening wild stocks of American ginseng (*Panax quinquefolius, **above***). Though it once flourished in woodlands throughout the eastern USA, several hundred years of collection have taken their toll, and it has become endangered in some areas. In some states, harvesters are required to scatter the seeds of mature plants they pick; in others collecting is totally banned.*

been successful in its struggle for existence and will impart its strength to the consumer. As a result, over-collection is endangering the wild resources.

In the west, the emphasis on curative medicine and the often theatrical results of new drugs long drew attention away from the potential of ginseng and other "harmonizers" in medicine, but interest has been growing for some time. A great deal of work, mainly in the former USSR but now in the USA too, suggests that these plants do improve overall fitness, increase physical performance, and aid mental alertness – especially under stress. The reason is still obscure but is thought to lie in the combination of chemicals involved. This makes replacement by synthesis difficult, if not impossible. *Eleutherococcus* is now prescribed generally in Russia for convalescence and chronic infection. Similar properties have been found in roseroot (*Sedum rosea*), and in the berries of *Schisandra* (*Schizandra*) *chinensis* and spp.

There are many other plants long known for their power to extend human performance without damaging health. Many people use herbal pick-me-ups such as sarsaparilla, sage, yarrow, mugwort and dandelion root. Onions were consumed on a prophylactic basis long after most of the traditions of herbalism died in the west. Capsules of garlic can now be purchased in any health food store, or raw cloves can be chewed. These pungent members of the lily family are held to reduce the chances of heart attack and stroke by lowering the level of cholesterol in the blood and reducing blood pressure, to stimulate insulin produc-

tion (thus lowering blood sugar levels), to regulate the menstrual cycle, promote wound healing, suppress allergy, counter asthma, even act as an aphrodisiac. Evening primrose oil, from the tall-spiked genus *Oenothera*, is gaining a reputation which may take it out of the class of panaceas such as ginseng or garlic and into mainstream medicine alongside foxglove and *Cinchona*. It is taken by multiple sclerosis sufferers and hyperactive children, for schizophrenia, migraine, and Parkinson's disease; to counter senile dementia, premenstrual tension, benign breast conditions, high blood pressure, and certain types of heart disease; even asthma and eczema. Medical opinion does not support everything said about evening primrose; there is guarded enthusiasm, but truly rigorous clinical trials are still needed.

Some plants long known in traditional medicine have now surfaced as scientifically based stimulants: kava kava from the Pacific islands; khat (*Catha edulis*), daily chewed in many Arab and other Muslim countries; cola nut (*Cola* spp.), caffeine-rich favourite West Indian masticatory and an ingredient of the original recipe of cola drinks; and milder tonics like jujube (*Ziziphus jujuba*), liquorice (*Glycyrrhiza glabra*), and senega (*Polygala senega*) have all been profitably investigated. One of the most notorious is the South American coca plant (*Erythroxylum coca*) which is widely cultivated in the Andean highlands. The conquistador Pizarro noted that the Incas chewed the leaves. Indeed the shrub was so highly esteemed that it was considered to be the property of the Incan royal family, and leaves were laid

The leaves of the coca plant (Erythroxylon coca) *have been seen on Bolivia's mountain terraces for centuries, since the days when the plant held religious significance at the Inca court. Cocada provides physical stamina and relief from discomfort enabling people to carry out hard labour on an inadequate diet. The leaves are a source of high income from illegal cocaine. The sight of government officials burning seized cocaine is not uncommon.*

in the graves of noblemen to serve them in the afterlife. Coca stimulates the brain and depresses hunger – the Spaniards made sure their Indian slaves had plenty of it. In Peru and Bolivia today, rural people endure the harsh conditions of their lives, carrying loads up and down mountain slopes and labouring at heavy work, by stopping periodically not for food but for the cocada or chewing pause. Coca was distributed to other tropical countries in the late nineteenth century, and the chewing habit has spread to Java, India and Sri Lanka.

The active ingredient, the alkaloid cocaine, was isolated from coca in the 1840s. It stimulates the nervous system, producing effects very similar to amphetamines but not nearly so long lasting. To maintain a "high", cocaine must be administered frequently. On withdrawal, a habitual cocaine user does not suffer the severe physical symptoms common with some hard drugs, but can become severely depressed. In the nineteenth century, cocaine was a common addition to elixirs and tonic powders. It was an original ingredient of Coca-Cola, though banned from the drink in 1904.

In 1884 a young colleague of Sigmund Freud discovered cocaine's amazing possibilities as a local anaesthetic, in which form it revolutionized surgery. But the potential for abuse was clear, so a substitute lacking the stimulant properties was sought. Analysis of cocaine led eventually to the synthesis of a non-addictive chemical substitute, lignocaine, which has largely replaced cocaine in anaesthesia (though cocaine is still used in eye, ear, nose and throat surgery).

Purges

Many plant species have been used as laxatives or purges.

The Egyptian *Ebers Papyrus* lists three of the classic cathartics then available – aloes, castor oil and senna – all three still used today. The oldest of all recorded purges, however, is rhubarb – not the familiar garden species, *Rheum rhaponticum*, but primarily *R. palmatum* and a handful of other species, mostly Chinese. Known for 5,000 years in China, rhubarb came to be valued so highly that in the sixteenth century the French paid ten times more for it than for cinnamon. The standard North American purge is cascara, obtained from the bark of a buckthorn tree, *Rhamnus purshiana*. An early Spanish missionary found the Native Americans in California and Oregon using it; he apparently gave it its popular name, cascara sagrada, "the holy bark", on account of its mild but highly efficient action. Nowadays its extreme bitterness is usually masked in chocolate-coated pills. Although the tree is cut down to remove the bark, the stump quickly produces new growth, so that – unlike the quinine trees – extinction does not follow collection. Its European cousin purging buckthorn *(Rhamnus cathartica)* had a similar traditional use, as evidenced by both its common and its scientific names. A Chinese species, *R. utilis*, is used to make Chinese green indigo for dyeing silk. Senna *(Senna alexandrina = Cassia senna)* contains glycosides in its leaves and pods and has an effect similar to that of cascara, stimulating peristalis. India exports about 6,000 tonnes a year.

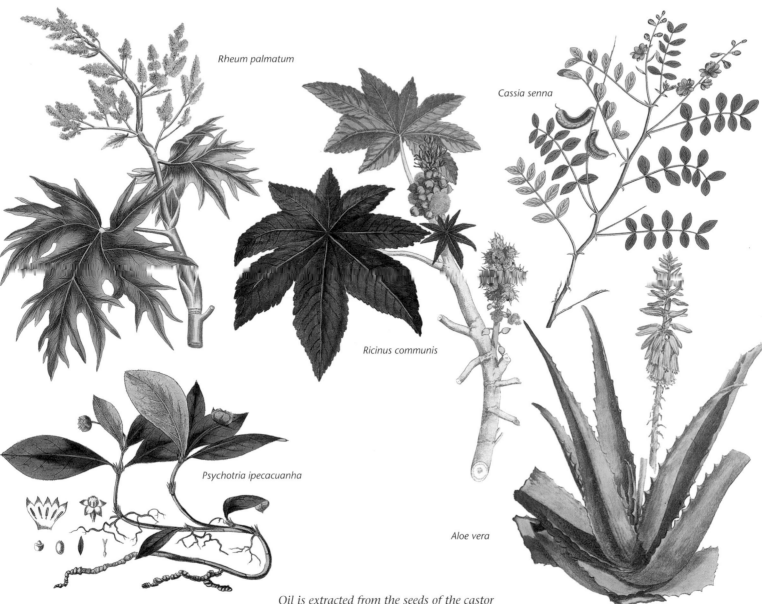

Rheum palmatum

Cassia senna

Ricinus communis

Psychotria ipecacuanha

Aloe vera

Ipecacuanha is the local name of a Brazilian jungle shrub, Psychotria ipecacuanha. *One of its alkaloid constituents, emetine, causes vomiting and may be administered in cases of poisoning. It is also useful for amoebic dysentery as well as bronchitis, coughs, whooping cough and croup.*

A powerful purge is obtained from the medicinal Chinese rhubarb (Rheum palmatum). Although there about 30 species of rhubarb, only a few are laxative. To be effective, plants have to be at least six years old; the dried roots, from which the bark is removed, contain the resinous purging material.

Oil is extracted from the seeds of the castor bean or castor oil tree (Ricinus communis), also called the palma Christi or wonder tree. It was known to Pliny and, indeed, the ancient Egyptians before him. Besides its value as a purge, castor oil is used in ointments and eye-drops and also has many industrial uses.

Though probably long used by the local Arabs, the merits of senna, obtained from both the fruits and leaves of a shrub belonging to the large genus Cassia, *were first made known by a famous Arabian doctor, Mesue, whom Harun al-Rashid invited to practise in Baghdad in the eighth century* AD. *It is now grown in India and the Sudan, and imports into the USA are measured in millions of kilos. It is primarily known as a laxative.*

As with rhubarb, so also with aloes: out of around 365 species only a handful were found to have the cathartic property, one a Mediterranean native, one from South Africa and, the most famous perhaps, from the island of Socotra south of Aden (east of the Horn of Africa) – famous because Alexander the Great, at the suggestion of Aristotle, sent experts there around 325 BC *to find out how this drastic purge was produced. Today, however, the flora of Socotra is critically endangered by rising herds of livestock; of the 850 plant species, about 260 grow nowhere else in the world, and about 50 are red-listed as threatened.*

These plant-based purges are ingredients in a good proportion of well over 700 laxatives advertised in the United States alone.

Traditional and modern

Of the known medicinal uses of plants, by far the largest proportion remains in traditional, folk or herbal medicine. At least 35,000 plants are used for medicine, including at least 10,000 in China and more than 7,500 in India. The more formal systems of traditional medicine also use many species – about 600 in traditional Chinese medicine (TCM), over 1,100 in Tibetan medicine, and as many as 1,400 in Ayurveda (notably in India and Sri Lanka). A great many more are in local use, and may not even be botanically identified. Three-quarters of the developing world's medical needs are met by herbal preparations – mostly crude medicines in the sense of being "off the tree", pulped or decocted and so on, rather than refined or synthesized.

In the last decade interest in traditional medicine has been renewed, and much is now being done worldwide to give it the respect it deserves – green medicine is being born again. The World Health Organization (WHO) is trying to incorporate traditional medical plant knowledge into health systems in developing countries, since western health methods are expensive and often inappropriate. The WHO has recognized that it will simply not be possible to replace traditional medicine with western health care for all people, at least in the immediate future. It also recognizes the important social function fulfilled by herbal-

ists and bonesetters. For the great majority of people, the local healer offers the only medicine available The hospital or clinic may be far away, and an alarming place, whereas traditional medicine maintains human dignity, fits in with the culture, and is familiar. So building a national health service out of a synthesis of traditional and modern medicine seems sensible – but it is a great challenge. Above all, the WHO seeks to promote plant medicine as chemistry, not magic, by applying scientific analysis to the herbal treatments used. This development has been greatly inspired by the example of China, where the practitioners of western medicine have been made to regard their traditional counterparts as equals, and today herbal medicine and acupuncture are being combined with what are considered the best western methods. In China, the Barefoot Doctor Programme, starting in the late 1960s, aimed to bring more effective medical services to the people. It was accompanied by a national inventory of TCM, and today there are 36 universities teaching traditional medicine in China.

The Vietnamese too have a remarkable record. Families are being encouraged to grow at least a dozen of the 58 recognized "family medicine" plants. The government buys most of its medicinal raw materials from these gardens, so that villages not only have health care at hand but an extra source of income. In recent years a massive drive to study local plant remedies produced no fewer than 40 new drugs for disorders as varied as high blood pressure, rheumatism, hepatitis, goitre, coughs, allergy and shock.

Other countries producing plant medicines include Bangladesh, Madagascar, Rwanda and Thailand.

In Africa, the daily lives of farmer, raiser of cattle and nomad depend on medicinal plants; in fact 95 percent of traditional drugs come from plants. Their healers serve a long apprenticeship and have great botanical knowledge of local plants, their uses and properties. All this knowledge is in grave danger of being lost, as increasing human and livestock populations put more pressure on natural vegetation.

The Islamic system of traditional medicine, common in Pakistan and generally on the Indian subcontinent, is, like western herbal medicine, rooted partly in the medicine of ancient Greece. It uses about 340 plant species, and is accepted as a state system alongside modern medicine. It has thousands of practitioners, qualified in medical colleges, and claims to reach all the rural people in Pakistan as well as many people in the cities. They work under a national council, which regulates and maintains standards. India too has a very elaborate system of traditional medicine with 900-plus hospitals and 14,000 dispensaries. However, the traditional doctors are sometimes regarded as underqualified and subordinate by those educated in western-style practice. Although integrated courses are taken by students, they are inclined to abandon the older Ayurvedic theory when they qualify.

But research into traditional plant drugs is increasing, even in India, as well as in Mexico, the Amazon region, and Ghana, where effective herbal remedies for guinea

Peat moss (Sphagnum), *known since ancient times, was widely used in World War I as a wound dressing, largely for its moisture-absorbing capacity. A Bronze Age skeleton found in Scotland appeared to have a chest wound with a large wad of* Sphagnum *applied to it. Its healing properties long remained unclear. Recently, analysis has revealed several associated micro-organisms, among them a* Penicillium *fungus, which might explain its apparently antibiotic action. The upper picture shows fruiting bodies; the lower a greatly magnified section to show the highly absorptive cellular structure.*

worm and *Herpes zoster* (incurable by other means) have been found. In time, research into traditional plant drugs may influence western medicine even more strongly. Many such drugs are on the brink of being brought into use – for dysentery, hepatitis, internal parasites, coughs and fevers, and skin diseases, possibly even leprosy.

The greatest virtue of good traditional medicines is their freedom from side effects, since they have been tested on so many generations. They are also usually cheap and locally available. Most important of all they are acceptable to the patients. However, outside their indigenous areas plant drugs need to be precisely identified, and the best and safest ways of using them worked out. If they are to have wider application, synthesis of the basic compounds may be desirable, partly because this often leads to improved action, but more importantly to prevent the over-collection of the plants concerned. Some are already endangered because of their local use – so much so that in Uttar Pradesh, for example, controls on collection of 13 plants important in Ayurvedic medicine have been imposed.

Scientific examination of the thousands of plants used traditionally is more logical as a first step than random plant screening, which has also been carried out up to a point but is costly and time-consuming. One problem with scientific assay of an indigenous medicine is that it is not necessarily one active ingredient in one plant that makes it "work". The wound-healing properties of *Spaghnum* moss, for example, are still unclear, after millennia of use. A modern case is the Ghanaian *Cryptolepis*

sanguinolenta, which reduces fever including malaria, cures urinary infections and is also a broad-spectrum antibiotic – yet none of the individual substances isolated from it is recognizably antibiotic. It may rather be the combination of several that is effective, and such synergism is common in herbal remedies. This notion runs contrary to western medicine which prefers one problem, one curative principle. Another difficulty is that some local practitioners are, not surprisingly, very wary of western ways. They do not want their remedies snapped up by international pharmaceutical companies, patented and sold back to them in expensive pill form.

There can be no doubt that there are many more drugs to be found in plants and that research into traditional remedies will continue to influence western medicine. Furthermore, chemists will never be able to synthesize all the secret substances of the plant kingdom. Thus the future contribution of green medicine strongly depends on the extent to which medicinal plants are conserved, or locally grown, and on the protection of wild environments. Research is urgent, since indiscriminate collection, either for local use or pharmaceutical exploitation, together with wholesale destruction of habitats, is seriously endangering not only the plants already in use but also those which may be useful in the future. Perhaps most frightening is the lack of knowledge of which medicinal plants are actually in danger. The People and Plants Programme of WWF has worked hard to shed light on this, and other plant conservation problems.

Prunus africana (Rosaceae) bark exploitation, Cameroon

Herbal medicines and conservation

Many countries use traditional plant, or plant-based, remedies to treat diseases and disorders, or to strengthen the body's resistance to infection, and in poor societies that have limited access to "conventional" (allopathic) medicines, plants help to keep people healthy, as they have done for thousands of years.

It is a surprise to many people that Europe also has a long tradition of using herbal medicines. The earliest "floras" were in fact "herbals" and depicted and described plants according to their medicinal properties, and the oldest botanic gardens, such as Pisa (1543), Padua (1545) and Florence (1550) in Italy, and Oxford (1621) and Chelsea Physic Garden (1673) in the UK, were founded as "physic" gardens, and used for growing medicinal plants and teaching about their uses.

Plants provide the predominant ingredients of medicines in most medical traditions. The number of species used worldwide may be 35,000–70,000 out of a total of more than 250,000. It has been estimated that 10,000–11,250 types of plants are used in China (over 95 percent of which are wild), 7,500 in India, over 2,200 in Mexico and 2,500 traditionally by North American Indians. The great majority of species of medicinal plants are used only in folk (orally transmitted) medicine, the more formal medical systems mostly utilizing rather fewer: 2,000–3,000 commonly in traditional Chinese medicine, about 1,100 in Tibetan medicine, and 1,250–1,400 in Ayurveda.

Although the various traditional medicine systems evolved separately, there have been connections between them. For example, in the eighth century AD, Trisong Detsen, King of Tibet, called a conference of medical experts from China, India, Nepal, Persia, Tibet and other parts to discuss the evolution of improved medicine, drawing on various traditions. This resulted in the development of Tibetan medicine, based on the pre-Buddhist Bon tradition of Tibet and incorporating elements from elsewhere, including from the medicine of ancient Greece.

Where the herbs used for medicine are gathered sustainably from the wild, or cultivated, the impact on wild supplies is not a cause for concern, but increasingly we see instances of over-collection, especially where the plants are the basis of commercial exploitation, often to supply lucrative export markets. Such markets have grown in recent years, as herbal medicines, often referred to as "complementary" or "alternative", have become more and more popular in the richer countries, especially in Europe and North America. This is where herbal medicines and conservation come together in problematic fashion. In Germany, for example, about half of all medicines used are based on plants, and in the UK the market in alternative medicines is worth at least £100 million a year.

The following two examples illustrate some of the conservation problems involved with medicinal plants.

The African cherry story

The story of African cherry (*Prunus africana*) shows what can happen when a local, traditional medicine is exploited for the export trade. Since most plant-based drugs are derived from plants collected from the wild, rather than from cultivated sources, this is a widespread problem.

The bark of African cherry is the major source of an extract which has been shown to be effective in treating benign prostate overgrowth, a condition that commonly afflicts older men. The bark is stripped from trees in montane forests in Cameroon, Zaire, Kenya and Madagascar, and the bark itself or processed extracts are exported to Europe, mainly to France and Italy, where it is used to prepare the drugs.

Although the tree is very robust and can in fact withstand removal of its bark, many trees are killed or cut down to supply the

Amchi *in the high mountains of Nepal holding specimens of* Delphinium brunonianum. *This member of the buttercup family grows on stony soils between 3,500 and 6,000 metres, and is rather rare in the area. It is used in traditional medicine to treat fever, headache, dysentery and appetite loss, as well as for body swelling and wounds. It is one of several that may be increasingly threatened by commercial exploitation, and associated over-collection.*

trade. In one six-year period, over 900 tonnes of bark were processed, which equates to an average of 35,000 trees debarked each year, probably affecting more than 6,300 hectares of forest annually. In Kenya, the tree may well be extinct within a few years if harvesting continues at current rates. The market is now thought to be worth nearly £140 million a year.

Local traditional healers and government officials in East and Central Africa are deeply concerned about the sustainability of this trade, as are conservation organizations, including WWF and Birdlife International. The pockets of rich forest where it grows are home to many rare plants and birds. WWF and others have researched the possibility of planting African cherry so that its bark can be harvested without disturbing the trees in their natural habitats.

Medicinal plants in Nepal

Medicinal plants play an important role in the health-care systems of traditional mountain societies. For example in the Himalayan region today, 70–80 percent of the rural population depend on traditional medicine for their primary health care.

The People and Plants programme has worked with local communities high in the mountains of north-west Nepal, where more than 200 species of plant are used medicinally by traditional healers *(amchis)*. Amchi medicine, also known as *Sowa Rigpa*, is based on the Tibetan medical tradition and uses a wide range of plants, as well as minerals and some animal parts. Most of the work centred on Shey Phoksundo National

Park, a remote and beautiful region far from the reach of conventional doctors.

Some of the medicinal species here are now threatened by over-harvesting from the wild, and the demand has increased due to commercial collecting for trade. One example is the perennial herb *Arnebia benthamii* (Boraginaceae) which grows on dry, open slopes, at between about 2,800 and 4,300 metres. It is used to treat blood disorders, high blood pressure, fever, coughs, and aches, and also as a source of purple dye.

Gentiana robusta (Gentianaceae) is another rare species, of open slopes and shrubby habitats, growing at between 3,500 and 4,000 metres. This is used for bile, liver, stomach and intestinal disorders, and to treat inflammation, food poisoning, swellings, and joint pains. The orchid *Dactylorhiza hatagirea* and *Nardostachys grandiflora* (Valerianaceae) are also used medicinally, and are locally threatened.

The project established a traditional health-care centre inside the National Park, in which the *amchis* produce medicines from locally harvested plants, and where they have experimented with cultivating medicinal species.

Some popular herbal remedies

Ginkgo *(Ginkgo biloba)* has been used for centuries in traditional Chinese medicine, and is increasingly popular in the west. It increases blood flow, and is used, among other things, to treat chilblains, and is also said to aid memory.

St John's wort *(Hypericum perforatum)* is also very popular. It is reputedly an anti-depressant.

Snake root *(Rauvolfia serpentina)* is collected mainly in India, Sri Lanka, Thailand and Bangladesh. It has been used for more than 4,000 years in Ayurvedic medicine to treat snake bites and mental illness. It is also effective for treating high blood pressure, menstrual problems, and as a tranquillizer.

Khella *(Ammi visnaga)*, native to southern Europe, contains chemicals used to treat angina, lower blood pressure, and to regulate abnormal heart rhythms, and also as a cure for asthma.

WARNING

Just because a drug is herbal in origin, it is not necessarily safe, especially if taken at the same time as prescribed conventional medicines. All medicines, whether plant-based or not, should only be taken after consultation with an experienced physician or expert. After all, the active compounds in medicinal plants are also chemicals, many of which are poisonous if taken at the wrong dose, or in the wrong combination.

CHAPTER EIGHT

Objects of Beauty

The story of the Garden of Eden expresses a universal characteristic of humanity – the love of natural beauty. From the Assyrian "Tree of Life" to the Greek acanthus columns, from the Minoan frescoes of lilies to the lotus of the east, our architecture, arts, artefacts and poetry all reflect the deep impression that flowers and foliage make on the human consciousness. Such appreciation is a luxury, yet given freedom from want and a little leisure, we are all gardeners at heart.

The gardens of the world are a treasure house of botanical skill, the creation of generations of individuals with a gift for beauty, rarity, and understanding of plants. The story of each flower, from its wild origin, through selection and cultivation, is unique and remarkable – though punctuated with some disastrous examples of over-collection and damage to the wild. Today, gardening is a more universal occupation than ever. And the further we are separated from the natural world by our cities and civilization, the more we fill our apartments and window boxes with plants.

Many of these indoor gardens depend heavily on tropical resources. Northern houseplants, which must live in shady, warm conditions year round, are often natives of tropical rainforests, where conditions are similar. But while we recreate tropical habitats in our homes, many of the plants we love are becoming endangered in the wild. The African violet *(Saintpaulia)* is as familiar to gardeners today as the tulip or daffodil. But in its native East Africa, the wild plant is now confined to a few isolated hill forests in Kenya and Tanzania, and many of these are threatened. In the deserts, particularly of

America, illegal collection and sale of cacti – so popular for sunny windowsills and desert gardens – has devastating effects on natural populations, which are slow to regenerate. The rarest orchid species are highly prized by collectors, and all around the world, wild flower populations are reduced by human impact and agriculture. Traditional garden varieties, too, are being lost, as gardening becomes big business and "new varieties" dominate the market.

When we look at buildings and artefacts of all kinds, it is clear that plants very early imprinted themselves on the minds of people all over the world. The example of temple column capitals is probably the one most often noted by the modern traveller. Papyrus decoration in Egypt, as at Karnak, is replaced with transformations of acanthus and similar foliages in the various Greek orders which were later taken over by the Romans.

Some of the oldest plant decoration is Assyrian and Babylonian, where plant motifs are often highly stylized, showing foliage and trees – the latter often symbolizing the Tree of Life. Occasionally these peoples made rosettes, the floral motif more widely used than any other down the ages. It was an integral part of roof decoration in Greece, as can be seen at the great theatre of Epidauros; and also features in the rose symbols of the Tudors and later. The rose can indeed be called the top decorative favourite throughout the world; lilies, pinks and carnations, daisies, violets, crocuses, narcissi, sunflowers, vines and of course leaves follow. Of most symbolic importance is the lotus: Egyptians, Assyrians, Persians, Indians and Chinese all depict it with reverence.

The tiled walls of this mosque at Isfahan are a spectacular example of the importance of floral decoration the world over.

Plants became important to many early civilizations both as symbols and ornamentation. Stylized leaves became habitual on the capitals of temple and palace columns in Egyptian, Greek and Roman architecture, as in this acanthus-leaf capital at Roman Ephesus in western Turkey, **left***.*

The Minoans in Crete were great devotees of floral decoration as demonstrated by this unique vase or crater, **below***, from the period 1900–1700 BC. Instead of the habitual painting on clay the flowers are three-dimensional; perhaps the potter was imitating stone and metal work, as suggested by remnants of chain hanging from the rim.*

One of the best sequences of vegetative decoration can be traced right through ancient Greek archaeology starting around 2000 BC when Minoan pottery from Crete began to exhibit its extraordinarily robust decoration, depicting such things as palm trees, irises, papyrus, and often seaweed-inspired motifs. The Minoans painted frescoes including beautiful depictions of lilies, irises and sea daffodil *(Pancratium maritimum)*. Some of their later vases carry appliqué flowers reminiscent of art nouveau. In mainland Greece floral decoration surfaces on pottery in the Archaic period (eighth to sixth centuries) and continues on incredibly detailed vases up to the Romano-Hellenistic period. Foliage represented, apart from acanthus, includes oak, olive, grapes and ivy, and there are also motifs based on the tiny seeds of annual weeds.

In many examples of ancient buildings it is almost impossible not to find plant motifs, whether one is in Italy, Greece, or India and the Far East. Persian flower decoration on tiles and ceramics is notable, while in India inlays of semi-precious stones in marble, as well as delicately carved reliefs, embellish the Taj Mahal (the most famous example). In the Chinese civilization, which matured so rapidly, accurate depictions of flowers appear on porcelain, and in external paintings which embellish almost every surface of places like the Summer Palace in Beijing, sometimes picking out work in carved relief. Weird objects of plant origin to be seen in some gardens are chairs and bizarre figures made of magnolia roots. In gardens also, doors and wall apertures are often in the shape of flowers or fruits, and one can see in the pebble mosaics of the Forbidden City's Imperial Garden all manner of flowers, fruits, and even flower arrangements.

European cathedrals and churches carry relief work of plant inspiration in stone outside and on woodwork within and, gradually, vegetation came to be used domestically on furniture, panelling and plasterwork. Some strange plants can be found by the discerning: there is the pea-pod decoration in Renaissance work, and stylized catkins decorate many a mantlepiece by Robert Adam. Their close resemblance to those of male *Garrya elliptica* poses a minor conundrum as the last of the Adam brothers died 12 years before the plant (found in 1827 by David Douglas in North America) flowered in Britain. These catkins must simply have been stylized hazel.

Brief mention must be made of the recurring theme in Persian carpets of a river in the shape of a cross dividing a garden scene into four parts: this represents the cosmic cross whose first reference is in the second chapter of Genesis – "And a river went out of Eden to water the garden; and from thence it was parted, and became into four heads." This symbol can be seen in much garden design, including the Tudor.

The ancient cultures that perhaps took plant forms in everyday objects most to their hearts were those of Central America and Peru. Stone and ceramic objects, some purely decorative and many utilitarian jugs and drinking containers, represent especially the widely used gourds, squashes and melons. The stylized "pineapple pot" has been made for centuries. Stone cacti appear to have been ritualistic.

Conventional flower painting of the Minoan civilization recently excavated at Thera on Santorini quite recognizably depicts sea daffodils, bulbous, autumn-flowering plants of Mediterranean beaches. The fresco, **above***, dates from about 1500 BC, shortly before the volcanic explosion which destroyed the colony. Interest in plants as decoration and a subject for painting was even older in China. Many buildings are ornamented with plant motifs, like this painted wooden relief,* **below***, on a Kunming temple – perhaps a tree peony (early favourite of Chinese gardeners) but possibly a water lily or lotus.*

In even the earliest paintings plants are very frequently depicted, either as symbols – the Madonna lily being a Christian symbol of the Virgin and of purity, for example – as part of a design, or as portraits for botanical purposes or, eventually, for their own beauty's sake. Plant forms exist in every department of architectural design and interior decoration, household objects, and in carpets, embroidery and wallpaper.

Styles of garden

Gardening traditions arose independently in eastern and western Asia, Egypt, Europe, South America and, probably, Polynesia. The first decorative gardens were certainly made in Egypt by 1500 BC; though the plants in them were utilitarian there were walls, regular beds and pools, and trees featured strongly. Not long after, the Egyptians were including plants grown for their beauty alone, such as the water lily. A little further east the Assyrians and Babylonians made great tree parks; they were partly for hunting in but also to create, as Francis Bacon put it, "A refreshment to the spirits of man." They built artificial hills called ziggurats to raise greenery above the featureless "land between the rivers"; the Babylonian "hanging gardens" of Nebuchadnezzar II were specifically made for the enjoyment of his Medean wife who longed for the wooded hills of her home.

In the Far East, the first Chinese pleasure gardens certainly existed by the fourth century BC. The Chinese were early devotees of the beauty of individual flowers, and there is a Chinese proverb which says, "If you have

two pennies left in the world, buy a loaf of bread with one and a lily with the other." Their gardens were very stylized – idealized landscapes where waterworn rocks symbolized those extraordinary pinnacular peaks which sometimes stud the natural landscape.

Japanese gardens, which date from about the third century AD, were far more abstract and with strong symbolic features. Both the Chinese and Japanese regarded (the latter still do) their gardens as a replacement for nature beyond, which was often too awe-inspiring to face and too vast to be disciplined. In India, enclosures and water tanks – both for freshness and for potentates to disport themselves – were combined with trees, and quite early on flowers became important.

Climate controlled the design of European gardens when the Dark Ages gave way to the Renaissance. In summer-hot Italy, trees, both naturally "formal" ones like cypresses and sculptured topiary specimens, made avenues and focal points in highly architectural, usually terraced gardens with water and fountains. In France and Germany, where weather was less predictable, people would walk further in great formal parks, from one viewpoint to the next, marvelling at huge expanses of water and grass. In the Mediterranean, the olive, with its combination of shade and life-supporting crop, is widely planted, and the Judas tree with its brilliant pink spring display, and the horse chestnut, were early encouraged for shade and flowers.

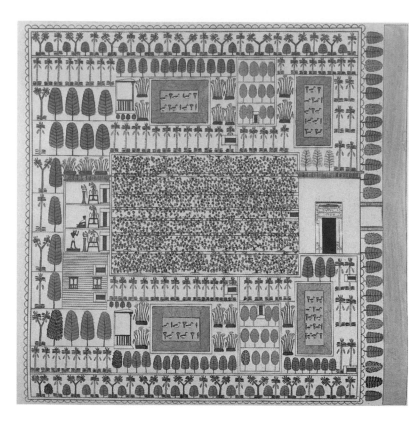

How plants travelled

As early as 1100 BC we find the King Tiglath-Pileser I writing, "Cedars and Box I have carried off from the countries I conquered, trees that none of the kings my forefathers possessed: these trees I have taken and planted in my own country, in the Parks of Assyria." Even earlier, Queen Hatsepshut of Egypt had collected incense trees from Somalia to decorate her temple at Deir-el-Bahri. But such collecting on an international scale was for a long time quite exceptional. Ornamental gardening started in principle with wild plants from the country concerned, and, in the case of Egypt, gardens would commonly contain date palms and sycamore fig trees, papyrus and the blue Nile water lily.

The earliest plants we now consider ornamentals were, with few exceptions, originally medicinal. This applies to the daylily (*Hemerocallis*), much prized for its herbal qualities in China, the prince's feather (*Celosia*) and the Syrian mallow (*Hibiscus syriacus*). Like the purgative rhubarbs from China, these travelled as articles of commerce on trade routes like the overland Silk Road or the ships of Arabian traders between China and Islam. The Syrian mallow had been established so long in western Asia that Linnaeus thought it a Syrian native, hence its name. From western Asia trade later continued westwards into Europe and brought such plants with it. One early introduction, the double annual balsam, seems not to have any medicinal properties; Chinese ladies used its petals to dye their fingernails. Some plants were traded in the reverse direction: for instance many narcissus species, mostly Mediterranean plants, came into Turkey, and the bunch-flowered *Narcissus tazetta* reached China over a thousand years ago.

Such early arrivals of new plants stimulated the intense botanical curiosity which

*In Egypt, **above left**, tall walls and plenty of shade-giving trees, with small, formal water pools, were the rule in 1400 BC. The same applied to many hot countries in later centuries: Islamic gardens were walled for the sake of the women, architecturally constructed with colonnades, formal pools and fountains, and planted largely with trees for shade and some bright areas of colour. Perhaps the finest example of a Moorish garden is at Granada in Spain.*

*In Kashmir, several well-preserved Mughal gardens like Nishat Bagh, **above right**, display their huge shade-producing oriental planes or chenars, regular flowerbeds filled mostly with annuals, formal pools, canals, and the sloped chiselled chadars or waterslides between different levels. An inscription at Shalimar garden reads, "If there be a Paradise on the face of the earth, it is here, it is here, it is here!"*

The basis of a Japanese garden – this is the West Garden of the Heian Shrine, Kyoto, **above** – was always an island in a lake, with a bridge for access; there might be artificial mountains. Later, planting was usually reduced to a few specially shaped trees among rocks, water or sometimes raked sand in "dry landscape", stepping stones and various architectural features.

In Britain, after stylization in the Italian manner of Tudor times, the garden became a place for perambulation. Shade was not essential so trees became decorative features or fringes to lawns of grass which the climate encouraged to stay green all year. Fine trees were available but the floral beauty to which we are accustomed today had to await the importation of exotics. Among these the bedding plants, annually renewed, give scope for perhaps the least natural form of gardening, like those which brighten the formally hedged, entirely symmetrical Sunken Garden at Hampton Court, **left**.

Gerard, in his Herbal *of 1597, was able to refer to 678 groups of plants, many of which comprised numerous species and varieties. The transition from the practical to the ornamental, however, was rapid: in 1629 John Parkinson was subtitling his equivalent book "A Garden of all sorts of Pleasant Flowers" – its title page is shown* **above** *– and his emphasis throughout is on their beauty and garden use.*

was a feature of the Renaissance. One of the great stimulants must have been the discovery and conquest of Central and South America by the Spaniards and Portuguese, which led to the importation to the Old World of many decorative plants like those now called French and African marigolds *(Tagetes)*, dahlias, passion flowers, the Jacobean lily *(Sprekelia)*, the tuberose and succulents like agave and prickly pear which are now naturalized features of Mediterranean countries.

All early movement of ornamental plants was between institutions such as the earliest botanical gardens (first Pisa founded around 1543, and Padua in 1545) or between enthusiastic amateurs. It became habitual for government scientific expeditions to carry a botanist – as did Captain Cook (his first and greatest being Sir Joseph Banks) – and to ask sea captains to bring plant treasures back from the countries they visited. Thus around the turn of the eighteenth/nineteenth centuries many camellias and the equally prized azaleas were brought to England by East Indiamen, some on their own account, some for plantsmen like Sir Abram and Lady Hume – one camellia became known as Lady Hume's Blush, while another, a very beautiful form of *Camellia reticulata*, commemorates the captain responsible, Captain Rawes; it is still in commerce.

Nurserymen began to sell ornamentals to paying customers in the seventeenth century, when trade became established between European and Turkish nurserymen mainly in the fleshy-rooted plants like anemones, buttercups, hyacinths and tulips which the Turks had so remarkably devel-

oped from wild species by selection. Horticulture did not become big business, however, until canals in the late eighteenth and railways in the nineteenth centuries eased transport problems. Then, when nurseries arose near all the larger towns, nurserymen began to send people abroad to collect new plants. This was first entirely a European business, with the British, Dutch, Belgians, and French most prominent, although paid collectors also set forth from the USA; and there were a number of free-lance collectors.

Most of the collectors were gardeners, not trained botanists, and few had ever been abroad before. It is remarkable how efficiently most of them survived in lands where they could not speak the language and where unimaginable hazards abound, quite apart from privations, and privateers, while on long ship-board journeys. They became devoted to their calling and it was unusual for them to retire while they could still work. Many died on the job, typically of dysentery and tropical fevers or, in the case of David Douglas, by falling into a pit-trap containing a wild bull.

Many houseplants are named after people – especially plant collectors and botanists: the great Joseph Banks (1743–1820) is commemorated in the genus *Banksia*; David Douglas in *Douglasia*, a small North American alpine, and the Douglas fir. This tree's scientific name, *Pseudotsuga douglasii*, was, however, later changed to *P. menziesii*, after an earlier botanist.

By the start of the nineteenth century the desire for ornamentals had spread to the middle classes. The first collectors went to

North America, for hardy plants were at first most in demand; but the rage for bedding plants resulted from exploration in the tropics, notably South America. This coincided with vast improvements in glasshouse design, heating methods and the quality of the glass itself. It became fashionable for the wealthy to build greenhouses and conservatories.

The London (later Royal) Horticultural Society, founded in 1804, also sent out collectors, of whose findings subscribing members got a share. Notable expeditions were those of David Douglas to western North America in 1825 and of Robert Fortune to China in 1842, but plant collecting remained almost entirely in the hands of nurserymen (a notable exception being Joseph Hooker's important expedition to Sikkim in 1847, which was mainly self-financed). The scale of business may be judged from the payment by Joseph Knight, who ran the Exotic Nurseries in Chelsea, London, of the then enormous sum of £1,500 to William Baxter for seeds he had sent during one year from Australia. Without the commercial propagation carried out by the nurseries the work of the collectors would not, of course, have been widely distributed. Apart from disease and physical hazards, there were often political difficulties for the collector to overcome. While Spain and Portugal owned South America it was almost impossible for plant collectors to obtain permission to enter the continent. Apart from a few Jesuit advisers in the eighteenth century, China was inaccessible to Europeans until 1840, while Japan was closed until 1860. A few plants had been obtained from all these countries by interested parties attached to trading missions, but no systematic collecting was possible.

The number of plant introductions gathered great momentum over the centuries. Taking Britain as the example – undoubtedly the country with the greatest long-term interest in plant acquisitions – the numbers of plants known to be introduced were 84 in the sixteenth century, 940 in the seventeenth, and 8,938 in the eighteenth. Hazards diminished, travelling became more reliable, trading in Spanish domains was permitted, and British colonial and political power increased, so that between 1731 and 1763 the tally of introductions doubled; and by the time Victoria was made Queen in 1831, over 13,000 exotic species were known in cultivation.

The domination of the nurseries in plant collecting diminished in this century with the reappearance of the eighteenth century-style syndicate, the number of subscribers increasing greatly to include growers, nurserymen and botanical institutions. The figure who most clearly demonstrated this was E. H. Wilson, who first opened the rich flora of western China to European and North American gardeners. His first two expeditions were financed by the great Veitch nursery in England, but two later ones, as well as later trips to Japan and Taiwan, were backed mainly by Professor Sargent at the Arnold Arboretum, USA. His successors such as George Forrest and Frank Kingdon-Ward were entirely financed by subscribers who, of course, received their dividends in the form of plants and seeds successfully brought home. The practice continues today.

The houseplant Sparmannia, *the house lime, was named, with misspelling, after a Swedish botanist, Anders Sparrman (1748–1820) who sailed with Captain Cook on his second voyage. Though an energetic collector, his only remarkable adventure seems to have been his return to his ship stark naked after being surprised by hostile natives while swimming from a Pacific island beach.*

The common corn poppy (Papaver rhoeas), **above**, *varies slightly in the wild. About 1880 a Surrey vicar found one with a white petal edge; by sowing its seed and selecting unusually coloured seedlings he finally achieved the multi-hued strain – called Shirley poppies,* **below**, *after the Reverend Wilks's home – which are easily grown annuals; modern strains can be either single- or double-flowered.*

Plants for pure pleasure

As the distinction between useful and decorative plants became blurred, plants were increasingly grown purely for their attractiveness and also oddity: succulents, for instance, were early favoured (and also easily maintained). Increased leisure and the ability to buy herbal remedies rather than having to grow them oneself were important reasons. As interest became focused on plants for their own sake, increased attention was given to mutations or "sports" which transformed well-known flowers (in the same way they transformed crops like cereals). The most pronounced mutation is the "doubling" of flowers, when an abnormal number of petals is produced or, as in the daisy family, showy ray florets take the place of flat, central disk florets. This genetic accident can be seen in the wild from time to time. In Mexico, for example, local populations of *Tagetes* have been noted containing a high proportion of doubles that would not disgrace the seed catalogues of modern French marigolds. Such mutations can be brought into cultivation and increased by seed or other means.

Parkinson depicts a range of double *Tagetes* which show what we now call French and the bigger African types, names which they acquired during their early peregrinations when people did not know their origin. In this daisy family, other annuals prized more in double than single forms are the Scotch marigolds or calendulas, asters and zinnias, while the perennial chrysanthemum from China and the double dahlias from Mexico are far preferred in general to their single forms. Single pinks are favoured but one cannot imagine their near relative the carnation as other than double; nor has the single stock much appeal. The multi-petalled rose is esteemed more widely than the single (even if there are some gardeners who prefer the single's natural grace to the cabbage-like propensity of the double). In the medieval garden the medicinal single peony must have looked attractive, but today's double garden forms (many acquired from China) are objects of surpassing beauty.

There are plenty of exceptions to this statement of course. The double tiger lily is a monstrous object and virtually no other double lilies exist. Perversely, the double forms of datura, trumpet-shaped like many lilies, are considered very beautiful. Some flowers seem to have a real propensity for "unnatural" forms. The chrysanthemum and dahlia now have so many that modern competitions are divided into scores of classes to accommodate them. Another example is the camellia, originally a single flower with a prominent central ring of stamens, which has among its huge array of cultivars (largely the products of gardening in antique China) semi-double and fully double forms, the latter either informal in petal arrangement or so geometrical they might have been drawn with a compass; others, known as anemone-centred, have a central cushion of smallish "petaloids" within ordinary petals.

Sometimes a flower becomes utterly transformed by a mutation. The old-fashioned lupin, only introduced from British Columbia in 1826 (by Douglas), had a lax spike of down-pointing flowers. A jobbing gardener, George Russell, began to specialize in the lupin. Among his seedlings he pinpointed a mutation that made the flowers horizontal and hence the spike compact. This he followed by breeding with existing colour forms and incorporating the yellow tree lupin in the parentage; but the lucky break was that mutation.

Mutations altering colour are quite frequent. The commonest by far is to a white or albino form, but others arise which, carefully selected, can produce a whole rainbow of colours in a species. The modern race of Shirley poppies with its wide colour range was raised, from about 1880, by the Reverend W. Wilks of Shirley, Surrey, who selected and reselected seedlings which arose from a single wild red poppy with a white petal edge.

Some of the earliest selections were made, incidentally, by artisan gardeners in the eighteenth and nineteenth centuries. These fanciers, or "florists", established certain stringent rules for their plants, and any seedlings that did not conform were ruthlessly disregarded. One early favourite was the pink, which was the hobby of the weavers of Paisley; other florists' flowers were auriculas, polyanthus, tulips, hyacinths and ranunculus – the last a wonderful flower now only available in limited mixtures, of which over a thousand named sorts were once recorded.

Such florist flowers, perennial plants like the herbaceous peony and shrubby ones like the tree peony or moutan are propagated vegetatively – by division, bulb offsets or by grafting in the case of the moutan. (The

Chrysanthemums have a very long history. The Chinese grew C. indicum with small yellow flowers about 500 BC and, about a thousand years ago, the white or purple-flowered C. morifolium. Accidental breeding between these two produced the ancestors of our many modern kinds, which include fully double forms much grown by exhibitors, and curiosities such as the "spoon-petalled" varieties old and new shown here.

*Single roses like the variable Eurasian dog rose (Rosa canina), **above**, are undoubtedly distant ancestors of the most popular type of modern rose, the hybrid tea – fully double and perfectly formed according to modern standards – like Chicago Peace, **below**. However, double roses of various forms arose early in the history of gardening by mutation and spontaneous cross-breeding, from Europe to China.*

The immensely popular, vigorous and easy-going Clematis jackmanii, **below**, *was created by the deliberate crossing of large-flowered* C. lanuginosa *from China and small-flowered Eurasian* C. viticella, **above**, *by the nurseryman G. Jackman in 1858. Further crosses using the hybrid as one parent, and also mutations from it, have given rise since then to a quantity of large-flowered garden varieties.*

latter was, incidentally, one of the earliest speciality plants – in China, around AD 700, a hundred ounces of gold might be paid for a new variety.) In such ways established cultivars would be maintained. Accidents of nature like albinos are propagated likewise. A classic example of such a unique freak is the contorted hazel known as Harry Lauder's walking stick *(Corylus avellana* "Contorta"*)*, spotted in a Gloucester hedgerow in 1863.

An even more peculiar accident (if not a specially beautiful one) is the small tree *Laburnocytisus adami*. This arose in a French nursery in 1826 when the ornamental broom *Cytisus purpureus* was being grafted onto common laburnum rootstock to give it more vigour. The *Cytisus* scion was accidentally knocked off except for a tiny piece. This resulted in what is called a chimaera, the broom tissue being entirely surrounded by laburnum tissue. The resulting tree randomly produces lilac-purple broom flowers, yellow laburnum flowers, or intermediate purplish-pink ones.

Deliberate breeding of garden plants was the next step after selection of seedlings. The first deliberate crossing of two different plants was in 1717, when Thomas Fairchild applied carnation pollen to the stigma of a sweet William (both are members of the genus *Dianthus*). The result, like a large double sweet William, was sterile so no more came of it. Amateurs later began to breed garden flowers for "endless interest and amusement" (to quote William Herbert, later Dean of Manchester) early in the nineteenth century. William Herbert specialized in gladioli, heaths and bulbs of the amaryllis family; contemporaries devoted themselves to

delphiniums, petunias, lobelias and verbenas. The ever-popular *Clematis jackmanii* was produced from deliberate crosses in Jackman's nursery in 1858, and the first hybrid lily, *Lilium parkmanii*, was bred in 1879.

Some notable garden plants we owe to the bees, which for instance in 1826, at the chateau of M. Soulange-Bodin near Paris, transferred pollen from a *Magnolia liliflora* to an *M. denudata* nearby and so gave rise to *M. soulangiana*, today the most popular magnolia of all.

Early in the century too the first rhododendron hybridizers set to work (the earliest recorded cross had been made in 1796). Enormous enthusiasm for these large-flowered shrubs or trees had arisen with the introduction of various species; the sendings of early expeditions such as Joseph Hooker's resulted in huge quantities of seeds which germinated like mustard and cress and were planted out in prodigal numbers, arguably ruining a number of British gardens in the process. Early in the twentieth century, with an enormous influx of Chinese species from Wilson, Forrest and Kingdon-Ward, promiscuous hybridization became a fashion especially among enthusiastic amateurs, and today there are many thousands of hybrids recorded.

The rhododendron species still exist alongside these, and one can find the parents of the innumerable fuchsia cultivars, but one will seek in vain for the parents of such popular plants as pelargoniums ("geraniums"), petunias, tuberous begonias and hippeastrums (amaryllis).

At first haphazard – as in crop plants – breeding became more precise when the

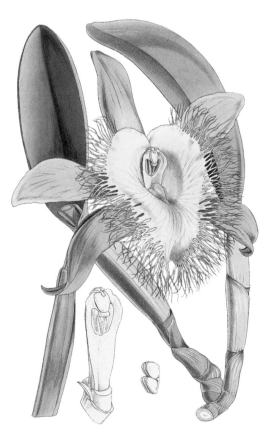

Cultivation in gardens and greenhouses allows hybridization of plants which never meet in nature. Among orchids up to four parent genera have been involved. The hybrid-genus Potinara, **above**, *is the result of crossing* Brassavola digbyana, **below**, *with an existing trigeneric hybrid called* Sophrolaeliocattleya, *earlier derived from species of* Sophronitis, **above right**, Laelia *and* Cattleya.

rules of inheritance became appreciated early in the twentieth century. A true blue rose has never appeared (though there are rumours that some purist breeder suppressed one, considering it quite "wrong" for a rose), but there are now red delphiniums (which a lot of people do not think quite proper either). In aiming for certain characters, like bigger sweet peas with more flowers on a stem, or yet further colour and shape variation in roses, there are often complaints that these habitually fragrant flowers are deprived of scent; but there always remain many cultivars with plenty of perfume. This does show, however, that even with scientific knowledge plant breeding can be a lottery, especially with flowers of such ancient and mixed parentage as the rose.

In very general terms one can say that hybridization promotes greater vigour, more variation in flower shape and colour, and more abundant flowering; with annuals or bedding plants it can ensure regular height. Often the wild parents are more graceful and elegant, but as in all things this is a matter of taste. At the present time, for instance, there is a definite swing away from the biggest dahlias in the world to smaller, daintier blooms, and a corresponding urge

in several specialist areas to keep to species rather than hybrids.

One curious result of present day universal travel is the possibility of breeding together plants that could never meet naturally. Among orchids, where this has been practised to the most extraordinary degree, there are now many trigeneric and a few quadrigeneric hybrids.

The tulip

Most important garden flowers, like the tulip, have a complex history due to their ancestry, human predilictions for their varying natural forms, and their resulting selection and breeding. Wild tulips grow in Europe, becoming more numerous to the south and east of the eastern Mediterranean and Turkey to Turkestan, with a handful in north-western China.

Though recognized as beautiful in the twelfth century, when Omar Khayyam referred to it, it was not until the sixteenth century that the tulip began to feature in Persian art. At that time Turkish rulers collected wild bulbs to decorate their gardens. There is a record of an order for 50,000 bulbs in 1574, and no less than half a million bulbs in 1579. This is one of the first examples of ruthless collecting from the wild and it is easy to see how ravages on this scale could render any common species extinct.

The original introductions to Europe came from Augerius de Busbecq, Emperor Ferdinand I's Ambassador to the Turkish potentate, Suleiman the Magnificent (one of the first to plant wild tulips in his gardens). De Busbecq sent seeds and bulbs to various people, among them being the Fuggers, the bankers of Augsburg; it was their plants of which the famous botanist Gesner published descriptions and illustrations in 1561. In his honour these first tulips were given the name *Tulipa gesneriana*, though this is now recognized only as the name of a group.

Seeds also reached the botanist and horticulturalist L'Ecluse (sometimes known as Clusius); he raised them successfully but asked so much for his bulbs that the best of them were stolen by a group of fanciers and are thought to have become the basis for the Dutch tulip industry, soon to give rise to that welter of senseless speculation known today as tulipomania (1634–37). One of the most extraordinary transactions in this mania is the recorded payment, for one single tulip bulb, of four oxen, eight pigs, twelve sheep, two loads of wheat, four loads of rye, two barrels of butter, a thousand pounds of cheese, two hogsheads of wine, four barrels of beer, a silver beaker, a suit of clothes, and a bed!

Though we cannot trace a precise botanical type of the ancestral garden tulip there is a possible clue. Some of the bulbs imported into Europe, now known as Neo-Tulipae, became naturalized, particularly

around Florence as did others at Aime, on the Little St Bernard Pass in Savoy. Perhaps a cartload of bulbs was accidentally jettisoned here, and some of their seedlings reverted to the appearance of the original species. At any rate, at Aime was found the plant now called *T. marjoletti*, which opens cream and then turns white, the margins of the pointed flower segments becoming edged with rose – the only white-flowered Neo-Tulipa known. It has the pointed segments that the Turkish cognoscenti admired.

Gerard said in 1597 that one might as well try to number grains of sand as to enumerate all the varieties of garden tulip, so one surmises that they were already fairly complex hybrids when they reached Europe. They were then further intermixed. Breeders searched for the black tulip, immortalized by Dumas: but as time went on interest became centred on tulips with coloured veinings, the plain ones being known as "breeders". The others had to become "broken" – with feathery markings – to be of any value. We now know that the "breaking" is due to a virus infection; this seems to have done the plants little harm and some early varieties such as "Keizerskroon", dating from 1680, and "Zomerschoon" are still in cultivation.

For long the tulip was a fancier's flower, and there were strict rules as to the form and patterning that were acceptable. But, as bedding out became more popular in the nineteenth century, there arose a demand for a bedding tulip. The "breeders" were self-coloured, but no attention had been paid to uniformity of height or of flowering times. These now became desirable qualities

and by 1899 tulip breeders were able to introduce the tall, sturdy race of Darwin tulips which were of level height and flowered simultaneously. These Darwins were in turn crossed with the dwarf Earl tulips, both single and double, which seem to have persisted unchanged since the seventeenth century, to give the mid-season race known as Mendels. Pointed petals, admired of the Turks but ousted by the seventeenth-century fashion for rounded ones, reappeared in lily-flowered varieties, while parrot tulips arose with deep-slashed, garishly mottled petals. Modern parrots have strong stems but are less fantastic than the originals.

Disappearing beauties

Like early medicinal plants which were wrested out of the ground for local trade use, ornamentals were treated in the same way. In places with a long tradition of ornamental gardening they were, however, propagated domestically in nurseries: the Chinese chrysanthemums, camellias and tree peonies are prime examples, where the interesting variants that appeared from seed were increased vegetatively and distributed.

Many of the new species sent back by British and European collectors much later on were in the form of seeds; there was no other way of doing it in many cases. However, things like cacti and other succulents, and bulbs, which could survive out of soil for fairly long periods, were brought back as plants or roots. It was all a matter of convenience, and no one thought twice about taking live plants out of the ground if that was easy.

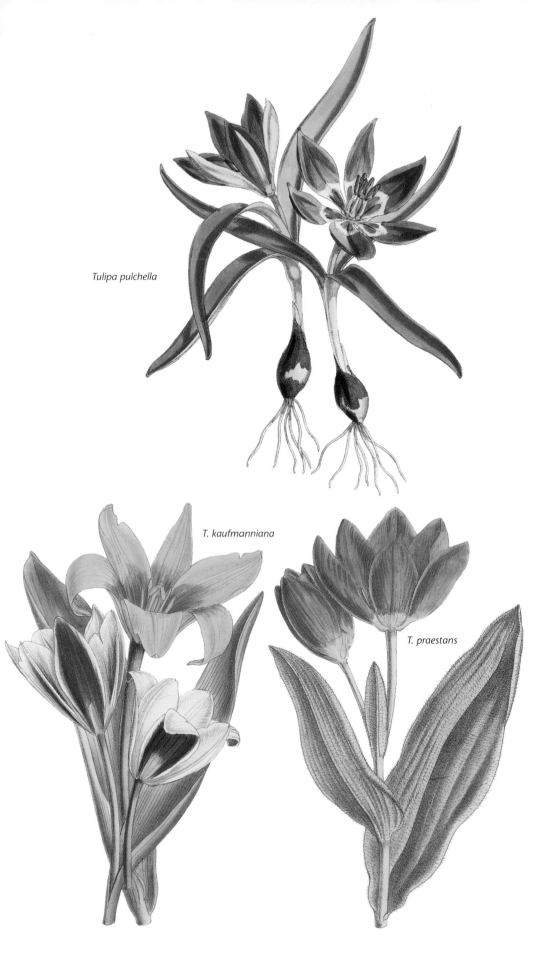

Tulipa pulchella

T. kaufmanniana

T. praestans

During the twentieth century something of a
tulip revolution occurred with the introduction
of previously unfamiliar species from
Turkestan. Tulipa kaufmanniana, *the water
lily tulip,* **below left**, *with cream flowers with
a red edge, flowers early in the season, and
there are now many more seedlings of it and
hybrids with other species. One of these is T.*
greigii, *with which it crosses naturally in the
wild. This tulip is noted for its purple-striped
leaves, and has a large red flower. Its garden
hybrids preserve the striped leaves, but the
flowers vary from red to yellow, often
strikingly marked within, and opening wide
in sunlight.*

T. fosterana *has enormous scarlet blooms,
usually opening in April: the most famous
variety is "Red Emperor". It has been crossed
with other tulips, giving rise to yellow, orange
and white varieties similar in form to the
original species. Between them these three
Asiatic tulips provide a host of low-growing,
early-flowering and mostly very long-lived
varieties before the main range of garden
tulips appear.*

*There are quantities of dwarf species
which the gardener can grow, mostly suited to
rock gardens, from the European fragrant
yellow* T. sylvestris *to the bunch-flowered* T.
turkestanica, *which may open its flowers in
February alongside the violet* T. pulchella,
top. *Others are the scarlet* T. linifolia *and* T.
hageri *and yellow* T. batalinii, *while the
season ends with the later-flowering* T.
sprengeri *with dull red flowers. Many other
wild species are grown by specialists. Most
need a summer "baking" without water.*
T. praestans, **below right**, *is remarkable for
bearing up to four bright scarlet flowers on
the one stem, though curiously enough it is
not the parent of the modern "multiflora"
races which are selections from old cottage
tulips with larger blooms.*

Paphiopedilum rothschildianum *is one of the striking slipper orchids with very side-spreading petals. Introduced in the 1880s, it lingered only a few years in the wild because of over-collection and was feared extinct. About 30 years ago it was rediscovered in South-east Asia, but continues to be in great danger from local orchid poachers. However, large numbers could be produced by either meristem culture or painstaking pollination and growing on of seedlings, as was demonstrated in 1983 at a Royal Horticultural Society's show when five plants were shown which had been grown from such "selfed" seed.* P. rothschildianum *may now be saved, in contrast to the similar* P. druryi *and* P. sukhakulii *which, stripped from the wild quite recently, are either extinct or severely threatened.*

Right into the nineteenth century this attitude did not change. Sir Joseph Hooker, travelling in India, remarks that *Vanda caerulea* is "The rarest and most beautiful of Indian orchids". Yet in a footnote he mentions that he collected "seven men's loads" of this plant (few of which reached England alive) in the Khasia Hills of Assam; he goes on to suggest that collectors with better facilities for getting the plants home "might easily clear from £2,000 to £3,000 in one season, by the sale of Khasia orchids" – and this is just what happened. Collectors swarmed into the forests where the orchids lived, as pinpointed by Hooker, and collected them by the ton, often cutting down the trees upon which they grew as the easiest way of doing so. Today this orchid is rare in its original habitat.

Orchids are among the plants to which most attention has recently been given to prevent such occurrences; as we have noted earlier, they are big business, and a rare orchid is worth a lot of money. Equally big is the business in cacti from the southern United States and Mexico especially, though *Rhipsalis*, succulent euphorbias and aloes from Africa are also endangered. The American deserts are being despoiled of their cacti in the most alarming way by commercial "haulers" who remove every kind of cactus by digging and often pulling out with a chain attached to the back of their truck (which does the plants little good), seeking "ordinary" species for garden use as well as rare specialities. The giant saguaro fetches $20 a foot – much more if it has a rare crested top.

It has been reported that 50,000 rainbow cacti (*Echinocereus* species) are dug out in Texas every *month*. The demand in the US is staggering: in one recent year nearly seven million cacti and succulents were *imported* (this includes a million from Mexico plus plants propagated in nurseries). Large numbers of the collected cacti are exported to fanciers abroad.

Apart from the local haulers, foreign collectors are equally to blame. A Japanese expedition to Cedros Island, off Baja California, left not one of the native succulents in situ. "Cactus study tours" originating in Europe, have been a cover for collection; a single party of German tourists was stripped of 3,600 specimens after returning from a trip to Mexico.

A few of the other ornamentals which are important in money terms are cycads (primitive, palm-like plants which grow extremely slowly), tree ferns and insectivorous plants. Apart from both American and Malayan pitcher plants (*Sarracenia* and *Nepenthes*) the insectivorous plant most at risk is the incredible Venus's flytrap: after heavy over-collection in the past it is now recovering well where protected, though drainage of its native marshes still represents a threat.

Large tracts of Portugal and northern Spain are devoid of wild narcissi, so great were the demands for bulbs on their wild populations early in the twentieth century. *Narcissus moschatus* and *N. cyclamineus* are very difficult to find, and *N. triandrus* ssp. *capax,* local to the island of Glenan in Brittany, is very rare. *Lilium pomponium* and *L. chalcedonicum* have nearly vanished.

International trade in many rare plants is controlled by the Convention on International Trade in Endangered Species of Wild Fauna and Flora (CITES), now implemented in over 80 countries. Originally designed to control trade in endangered wildlife products like rhino horn and the skins of jaguars and ocelots, it now covers about 5,000 species of animal and about 28,000 species of plants. The species are grouped according to how threatened they are by international trade. They include some whole groups, such as primates, cetaceans (whales, dolphins and porpoises), sea turtles, parrots, corals, cacti, and orchids.

CITES has, however, proved difficult to enforce for plants. The trade in wild-collected specimens is highly damaging but it is still small in comparison with the legitimate trade in propagated plants. So it is easy to conceal rare wild-collected plants and evade the laws. Perhaps a more realistic means of approach is to persuade hobbyists and collectors only to buy plants they know have been propagated artificially. Succulent plants societies in the UK have worked hard to outlaw advertising of wild plants for sale in their journals and to persuade their members to follow a code of conduct. After all it is far more rewarding to grow your own plants from seed and win prizes with them than buy in mature specimens, even though some of them such as cacti, cycads and tree ferns, for instance, are notoriously slow growers.

Houseplants are a good example of a group highly important to commerce where stripping from the wild has not taken precedence over propagation. True, most are easily increased; but one might argue that it takes almost as long to grow a showy

bromeliad from seed or offsets as an orchid from very similar treetop habitats.

Fortunately there is a brighter side to these problems. Nature is not always as efficient as she might be. The Monterey pine *(Pinus radiata)*, the flamboyant or flame tree *(Delonix regia)*, the Philippine jade vine *(Strongylodon macrobotrys)* and the horse chestnut *(Aesculus hippocastanum)* are near extinction in the wild through little fault of mankind's. The flamboyant, indeed, is only known from a very limited area in Madagascar. The maidenhair tree *(Ginkgo biloba)*, *Primula sinensis* and *Malus spectabilis* are virtually unknown as wild plants and survive mainly in cultivation; the ginkgo, preserved around Chinese temples, is now grown as a street tree in many countries. Wild ginkgo trees are still found in a few subtropical forests of central China, though the status of these is somewhat disputed. The dawn redwood *(Metasequoia glyptostroboides)* was only known from fossils when discovered in 1941 by a Chinese botanist; what is more, the few trees he found were being cut down for timber (more have in fact been found since). Seeds collected were freely distributed and grew readily into fast-growing trees.

The unique trailing succulent burro's tail *(Sedum morganianum)* was spotted in a Mexican market: it has never been found in the wild. The popular houseplant *Pilea cadierei*, known as aluminium plant, was found just once in an Indo-Chinese jungle and never since. Easily propagated, both are now widespread. In the future, establishment in cultivation may well be as important as a few heavily protected plants in the wild.

Narcissus moschatus *is distinctive with its creamy white flowers, though now regarded as one of the many natural variants of the Lent lily* (N. pseudonarcissus). *Native to the Spanish Pyrenees, its existence in the wild is now open to doubt; like most of the species of daffodils in Spain and Portugal, it was heavily over-collected in the early twentieth century. Bulb merchants paid local inhabitants for narcissus bulbs by the sackful, and the plants were eradicated over large tracts of country.*

Modern times

Distinct groups of plants, like bulbs or shrubs, and individual plant genera dominate much modern gardening. In Britain, and much of Europe, the most popular plant is probably the rose. This shrub, with species from most of the temperate world, was a Roman favourite. By that time the rose was already multi-petalled and its long lineage continued unabated, with many species entering in, until the introduction around 1789 of Chinese garden varieties that flowered summer-long – "perpetual" in nurseryman's parlance. These derivatives of *Rosa chinensis* (a species not, incidentally, discovered by Europeans in the wild until 1885) made all the difference to the garden value of roses today. Rhododendrons and azaleas, often accompanied by camellias, are predominant on both sides of the Atlantic. Hobby flowers once thought of only for the rich, but now very much more universal, are the orchids, grown in most countries under glass or in the open according to climate. More than any other plants, orchids are big business in money terms.

These and many, many others, have become the subjects of specialized societies – the list is endless. It would seem that whatever one's interest, there is a society to promote it. In Britain, the National

One of the important gardening developments of temperate countries since the late nineteenth century has been that of woodland gardens where exotic plants are sheltered by native trees, or at least the naturalized introductions of earlier generations.

Leonardslee in Sussex, England, **below,** *set on steep hillsides around old hammerponds, also exemplifies the specialization of many present day gardens, being almost entirely planted with rhododendrons, azaleas and camellias.*

A fine example of a modern speciality garden, Monaco's Jardin Exotique, is devoted to cacti and other succulents culled from the drier regions of the world. In some places, like California, climate encourages the cultivation of such plants, which are less water demanding than more orthodox garden plants.

Council for Conservation of Plants and Gardens has developed the scheme of National Collections of well over a hundred garden genera which have numerous cultivars, so that these will be preserved for posterity. Public, private, national and botanic gardens are all involved.

There are societies too for bonsai, that ancient art of dwarfing trees, and for decorating with cut flowers, both stemming from oriental ritual traditions. Virtually every specialist plant society results in competition at shows and equally in speciality gardens; some people concentrate entirely on, for example, roses, or alpines, or on bedding plants for a summer spectacle. Some connoisseurs search for forgotten cultivars from older days.

Climate is of course a controlling factor. In Britain and northern USA, there can be marvellous rock gardens and woodland plantings of hardy trees and shrubs, but in California and South Africa the displays are more likely to be of succulents. The Mediterranean-climate garden always seems to contain bougainvillea, hibiscus, mimosa and climbing pelargoniums, and all round the tropics, public gardens especially display what is fast becoming a rather standardized range of colourful exotics – hibiscus and bougainvillea again, with cassia and flamboyant, allamanda and solandra, fragrant frangipani, lantana and datura; some of striking form like primeval cycads, palms, and the fan-shaped traveller's tree; others with coloured foliage like croton, Joseph's coat and dracaenas.

Those who cannot grow exotics outside, cultivate houseplants, and these – largely tropical foliage plants – have become a huge trade especially in Europe and the United States. The houseplant fashion reflects the deep-felt internal desire people have to tend something green and living, to get their fingers into soil (even if it is usually a proprietary man-made "mix"), when they have perforce to live without a garden in apartments and flats in cities.

There are few parts of the world, even in poor regions, where you cannot find plants grown ornamentally in some way. Ruskin aptly wrote, "flowers seem intended for the solace of ordinary humanity."

CHAPTER NINE

Plants and Society

It often comes as a surprise to sophisticated citizens of developed countries to realize just how much they depend on plants. But it is probably even more astonishing to discover that in many parts of the world whole cultures rely on plants, and their own skills with them, to meet every smallest need. Altogether, millions of people live this way, from the marsh Arabs of Iraq to the Amazonian Indians, from the Bushmen of the Kalahari and the Australian Aborigines to tribal peoples of New Guinea and Indonesia.

Enormous contrasts are explored in this chapter: on the one hand, "primitive" societies based on close involvement with natural vegetation, total awareness of its significance, intimate knowledge of its many uses; on the other, societies which have largely dispensed with natural plant cover and, for better or worse, transformed the land for human domination. Very different from either are cultures where a single kind of plant, with multiple benefits, assumes a unique symbolic and practical value.

These cultural differences shape people's daily lives; more importantly, they shape attitudes and knowledge. The suburbanite tackling a breakfast of cornflakes, sugar, toast, marmalade and coffee is hardly, if at all, aware of consuming plants (and tropical forest plants at that); but indigenous people are in direct daily contact with their green life-support systems and thus possess the most enormous store of knowledge about them – knowledge of which we shall be in ever-increasing need in the future. A recent survey of plants used by South American Indians, for example, turned up over 1,000

with potential economic value, as yet wholly unused by the west.

The Surinamese say, "In the jungle the Indian knows everything." The tribal peoples of the tropics are indeed intimate with their surroundings, able to live in harmony with their environment, and enviably managing their rich but fragile forests on a sustained yield basis. Until a hundred years ago, the Native American Indians, too, lived this way. Comparing their culture (using British Columbia as an example) with that of present day Amazonian peoples reveals, in both, an amazingly sophisticated technology. Each shows an intensive knowledge of every plant, its seasons, uses, and preparation. Each shows, too, a respect for the lives, the "souls" even, of the plants. "Look at me friend", prays a woman to the cedar she is about to cut down, "I come to ask you for your dress – for you have come to take pity on us; for there is nothing for which you cannot be used."

Native Americans of British Columbia

For many centuries before the arrival of European settlers, the Native American Indians trod lightly across the land, leaving it little altered. These resourceful people were able to find a solution for nearly every problem and, ingeniously, a use for almost any material, from their green surroundings. Sometimes one plant, like the western red cedar *(Thuja plicata)*, served many purposes; in this case for canoes, rope, clothing and blankets (from the bark). The range of indigenous plants used was large and

Bamboo forest, Anji County, Zhejiang Province, China. In the foothills of the mountains here, some 175 km south-west of Shanghai, are beautiful forests of mosu bamboo (Phyllostachys pubescens). These are cropped sustainably for their culms.

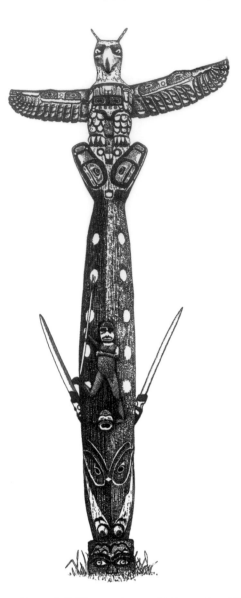

Above left *The great carved totem poles still stand in many places, silent symbols of one-time tribal coherence. These statuesque poles, recording family history, were usually made of western red cedar* (Thuja plicata), *a tree up to 70 m tall with strong, light, rot-resistant timber, quite easily worked.*

Above right *Babies were often back-packed in carriers, made from the pliable stems of the vine maple* (Acer circinatum, **above centre**) *or its relations – as were snowshoes and a range of openwork baskets. To mop up urine, nappies or diapers would be provided by shredded cedar bark, cat-tail, seed-fluff or rotten aspen wood. Absorbent branching lichens and* Sphagnum *moss were also used, the latter widely known also for dressing wounds; while wormwood* (Artemisia) *could heal nappy rash.*

included, for example, pinegrass/timber-grass *(Calamagrostis rubescens)*, Indian hemp *(Apocynum cannabinum)*, western service-berry *(Amelanchier alnifolia)* (berries for food), and snow buckwheat *(Eriogonum niveum)* (to treat infected cuts).

Among its myriad applications, red cedar also supplied housing – boards and planks for walls, bark for the roof. Cat-tail *(Typha)* yielded mats of braided leaves for insulation or for bedding, and offered its fluffy seeds to stuff mattresses and pillows, and young fruiting heads, sprouts and roots as food. Spruce *(Picea)*, fir *(Abies)*, ferns, black cottonwood *(Populus balsamifera)*, fireweed/rosebay willowherb *(Epilobium angus-tifolium)* – all contributed branches, fronds or seeds for warmth and comfort. In the hot summers of the interior, teepees framed of light branches and bark strips were preferred, or huts thatched with tule sedge *(Scirpus spp.)*.

The fire-lighting skills of Native American Indians are celebrated in folklore. These tribes used many fuelwoods, highly prizing the bark of Douglas fir *(Pseudotsuga menziesii)*, and other resinous conifers; for friction drills, they chose lodgepole *(Pinus contorta)* or Ponderosa pines *(P. ponderosa)*, and Pacific willow *(Salix lasiandra)*; fungi, cedar bark and sagebrush *(Artemisia triden-*

tata) served as fire carriers between camps, rotten willow roots as firelighters.

For weaving of clothing and mats, the versatile red cedar once more provided, yielding some of the best fibres. Leaves and stems of grasses and grass-like plants were woven too, on simple looms, or pounded into strands and then twined together. Sagebrush was used for moccasin insoles and tule, stitched with hardwood needles, for soft fabrics.

A Shalisham speciality was coiled basketry, in which spirals of fibrous materials were sewn together. These, and the woven baskets, were often decorated by intermingling strips of naturally or artificially coloured leavesor bark. Bitter cherry's *(Prunus emar-ginata)* tough, shiny, reddish-purple bark was often used in this way, sometimes dyed black. Some of these containers were so tightly woven that they were watertight. So too were the elegantly decorated red cedar boxes used for storing foods or to boil or steam them. The beauty of these Indian designs and artefacts has not been entirely lost, but passed into the modern tourist and craft industries. Red cedar also made the large (up to 60-seater), ocean-going canoes. The trunks were hollowed out in age-old dugout fashion. For light canoes, the river people (immortalized in Longfellow's *Song* of Hiawatha) looked to willow or cedar for flexible frames, to birch bark for the one-piece skin, and to conifer resin or pitch to caulk any cracks or seams. Some builders were so skilful that their canoes could be dismantled for portage.

Indians of the Amazonian jungle

The once numerous tribes of central South America inhabit the richest and most diverse forests in the world. Their skills echo those of the colder north, but with a completely different and much wider range of species. In every chapter of this book, plants from these jungles feature – food crops and spices, oils and fibres, medicines and contraceptives, orchids and houseplants. And in most cases, their "discovery" arose from local peoples' use of the plants concerned. Such contacts, however, have only brushed the surface of the Indians' knowledge of the forests. In their daily lives, plants provide everything.

The jungle has many discomforts, from parasites to vast numbers of irritating insects – tiny midges, mites that burrow into the skin, giant cockroaches that swarm round habitations; many plants are used to counter these pests, and the illnesses some of them may bring. Rainstorms have leached

virtually all minerals out of the soil, so the area has no salt; a brownish salt substitute is sometimes made by boiling up the ashes of certain trees, especially palms, to evaporation point. The fire is started with a drill, and rapidly boosted by adding lumps of rubber from latex-bearing trees, or some other resinous material.

Many of these peoples wear few or no clothes. Women may have a band of bark cloth, men a penis sheath woven from palm leaflets. Without clothes, personal ornaments take on great importance. The black, shiny nuts of *Astrocaryum* palms are widely carved into beads and earrings; beads are also made of fragrant woods such as the incense-scented pau santo ("holy twig") (*Kielmeyera coriacea*). For personal care, soaps and shampoos are made from several plants, notably several species of *Sapindus*; parasitic worms and mites are dealt with by smearing the skin with rubber latex, which suffocates the parasites. The Xingu Indians know a wild grass that is used as razor blades, and the Tapirape tribe used the crested fruits of a *Streptogyne* grass as tweezers to pluck hair.

As in the north, twine and ropes are vital. Various thicknesses are obtained from fibres, typically extracted by soaking leaves and picking out the strands. The common palm *Mauritia flexuosa* is a main source of cord,

Only a handful of the plant-based artefacts of these sophisticated people can be mentioned here, but this quotation from Plants in British Columbian Indian Technology *by Nancy J. Turner lists other uses of red cedar wood by coastal tribes – "Dishes, arrow shafts, harpoon shafts, spear poles, barbecuing sticks, fish spreaders and hangers, dipnet hooks, fish clubs, masks, rattles, benches, cradles, coffins, herring rakes, canoe bailers, ceremonial drums, combs, fishing floats, berry drying racks and frames, fish weirs, spirit whistles and paddles".*

Ritual objects are typically of wood like the North American shaman's rattle, **above left**, *in the shape of a bird with a human figure and a frog, and the Pueblo Indian kachina doll simulating a supernatural being,* **above**.

The orange dye from the seeds of Bixa orellana, **below**, *known variously as urucu, achioto, annato and by other names, is widely used to colour thread, ceramics implements and weapons, but most of all to paint the body and hair. Colorado Indians use it to colour the hair and, applied thickly, to create helmet-shaped hair styles,* **above**. *It gives magical protection from the spirits of the forest; its oil also acts as a mosquito repellent. The xagua or tapurita* (Genipa americana), *whose sap becomes dark blue after exposure to the air, is widely used both for dyeing and body painting.*

and other palms such as *Astrocaryum* and *Bactris* species provide the fibres used for sleeping hammocks, nets, and fish traps. The familiar raffia comes from several species of *Raphia*, especially *R. farinifera*. Many jungle Indians use fibres from the epiphytic bromeliads which cluster on the trees; in the uplands, the similar rosetted leaves of agaves are rotted in water to expose the fibres. Agave leaves usually have a fine hard point at the end and with care this can be extracted complete with the internal fibre leading to it, providing a needle and thread in one.

The jungle arsenal holds bows and arrows, spears, clubs and blowpipes. The most remarkable is the blowpipe – a tube about two metres long, bored from a palm stem – from, in the upper Amazon, the paxiuba palm *(Iriartea ventricosa)*. In Guyana, blowpipes have an inner barrel for extra accuracy. The outer tube is made from a palm and the barrel is formed by the stem of the rare bamboo *Arthrostylidium schomburgkii*. These components do not grow together: the bamboo is harvested only by the Machiritare tribe of the Pakaraima Range; the palm by lowland Indians. Other tribes again hollow the palm stems and fit the inner barrels to them, after which they are traded back for specialities, such as cassava graters.

The dart, a sliver of wood perhaps 20 cm long, is fitted at the base with a wad of cottony material from the fruit of *Bombax globosum* or *Eriodendrum samauma*; the sharpened tip is usually dipped in curare. The Indians have great knowledge of poisons for many different types of hunting: an ethnobotanist recently discovered a

potential heart-surgery drug among the forty or more different fish poisons used by Indians of Guyana; the toxin stunned, but did not kill the fish, and on investigation proved to be temporarily stopping the heart without killing the organ.

The lives of these forest Indians are intimately linked with the jungle and all its plants. Perhaps they should be recognized as its best guardians. One group, the Kuna of Panama (kuna is a local species of palm) have taken the bull by the horns and initiated a rainforest reserve in their own area – understanding that to preserve their culture, they need to preserve the forest, and that to achieve this, the best hope is a national park, partly open to research and tourism, the rest left wild.

One plant, many uses

Some peoples have depended so greatly upon a single genus, even a single species, for their daily needs that these uniquely useful plants have acquired a special status, as symbols of the culture they serve. Such is the date palm *(Phoenix dactylifera)*, grown since time immemorial in India, westwards to Mesopotamia, Palestinian lands, and North Africa, and most important in Egypt until very recent times. Other palms, too, have influenced whole cultures, from the coconut *(Cocos nucifera)* to the nipa palm *(Nypa fruticans)*. But the date has the longest and most illustrious history of all. There are now large plantations in Arizona and southern California. The fruits can grow to 12–15 cm long. They are classified into three types: soft, to be eaten fresh and ripe,

Most houses are small, their frames made of relatively light poles lashed together and a thatch of palm leaves covering the roof and walls. Some, such as those of the Yanomami, illustrated, are communal and may be very large: up to 30 m long, 20 m wide and 8 m high, with slats of bark, often painted with designs, for the walls. In the Upper Xingu, roof and wall curve continuously, below, like an aircraft hangar: these are made of lashed poles and are thatched with grass.

and containing only a little sugar; semi-dry, which are sun-dried and packed loosely, keeping a long time without fermentation; and dry, smaller dates containing a high proportion of sugar which, after sun-drying, keep indefinitely.

This last type, particularly, is consumed by the Arab people. Desert nomads find them especially valuable for their long-keeping, depending on their store of dates to supplement their standard diet of camel's milk. The dry dates are softened by soaking in water: rich in sugars and starches (up to 70 percent), they also contain fats and essential protein. Dates are used in confectionery too, in preparing pastries, and as an alternative to honey. The nuts are roasted as a coffee substitute, and the ground seeds are also used as camel fodder. Nomads even sometimes ate the growing tips as "palm cabbage"; but this destroys the tree.

Arabs have a saying that there are as many uses for the date palm as there are days of the year; the Babylonians referred to

360 uses – a mystical number. Mysticism aside, it appears that some 800 uses have actually been recorded, many of them involving the wood, which is used for building huts and houses, and for any number of implements which can be cut and carved from it. The fronds are used, entire, for thatching; individual leaflets and the long, hard leaf stems provide fibre, for ropes, cords, and baskets; stem fibre mixed with camel hair is woven into cloth, especially for travelling tents; mats are woven from the leaflets or their fibres.

The date palm is dioecious – the male and female plants are separate, so that the flowers of each sex are on separate trees. The knowledge of this is ancient; there are cuneiform texts of around 2300 BC from Ur, recording the necessity to carry out pollination. The traditional method is to cut male flower clusters when the pollen is ripe, climb each female tree, and tie a piece of the male cluster among its flowers. Muhammad taught his followers: "Honour

An English traveller wrote in the 1860s "What would a poor Chinaman do without the bamboo? Independently of its use as a food, it provides him with the thatch that covers his house, the mat on which he sleeps, the cup from which he drinks, and the chopsticks with which he eats. He irrigates his fields by means of a bamboo pipe; his harvest is gathered in with a bamboo rake; his grain is sifted through a bamboo sieve, and carried away in a bamboo basket. The mast of his junk is bamboo; so is the pole of his cart." He ends sombrely, "He is flogged with a bamboo cane, tortured with bamboo stakes, and finally strangled with a bamboo rope."

The giant bamboo, Dendrocalamus giganteus, *is still used for rafts in shallow turbulent rivers; while* Phyllostachys pubescens, *another tall species, has many uses, from scaffolding to furniture, and even for cables. The stems are split and twisted together and, in lines up to 500 m long, used by teams of men to tow junks up the rapids of the Yangtze gorges. Cables 65 cm in circumference are recorded as supporting a bridge nearly 250 m long and 3 m wide.*

In Nepal, bamboo is used for baskets, **left***; for matting to protect livestock,* **centre***; and for the Pakhola Bridge,* **right**.

the Date Palm, for she is your aunt on your father's side; she is made of the same stuff as Adam, and she is the only tree that is artificially fructified."

People have found even more uses for the ubiquitous bamboos of the Orient than for the date palm. First, many bamboo species provide food: the young growing tips of new culms (stems) are highly prized, especially those that are not bitter; the larger bamboos may have tips many centimetres across. Secondly, bamboo has special values as a material: light, but very tough indeed, with enormous tensile strength, the biggest kinds are virtually rigid, the thinner ones flexible. It can be split easily, but only in one direction. If heated, it can be bent, and on cooling retains any shape thus created. Artefacts can be made from the entire culm, reduced into sections where necessary (as for cylindrical containers), or by carving the wood into any shape or form; the split culm can usually be bent and woven, or reduced to strips of fibres of varying thickness, even to thread-like proportions. The long, papery leaf-like sheaths formed at every joint on the noded culm are used to make hats, fans and roofs, and as food wrapping. A few species are currently used for paper-making and also to form window panes.

Today, visitors to China and Japan will see tall buildings being erected with

bamboo scaffolding (using some of the vigorous species that can reach 25 metres tall); garden buildings made entirely of bamboo; and, in wood-framed houses, bamboo serving as walls, roofs, window lattices, in garden gates and doors, and as fences both ornamental and utilitarian. Bamboo supplies many surprising products, too, including ski sticks, vaulting poles and shuffleboard cues. Thomas Edison even used bamboo fibres as filaments in his early electric light bulbs before tungsten was available, and the plant once supplied gramophone needles!

Bamboo is one of the four noble Chinese plants (the others are the orchid, chrysanthemum and plum tree). Perhaps its most ancient association is with writing. The first Chinese books were handwritten, using a bamboo brush, on strips of bamboo lined with silk. It is also the subject of innumerable paintings, where in earlier times it served a moral purpose, to symbolize straightness. The brushwork needed to paint bamboo – there are a dozen prescribed basic strokes – is intimately related to calligraphy in both China and Japan. Bamboo remains integral to the cultures of these countries at every possible level: as the Chinese poet Pou Sou-tung wrote, 800 years ago, "Without bamboo, we lose serenity and culture itself."

Modern times – crops versus vegetation

Modern commercial agriculture has a very different attitude to plants from that of the cultures we have just examined. Instead of living off the natural vegetation in harmony with its cycles, agricultural peoples remove the original plant cover, and replace it with their own. Widespread cutting of trees and pasturing of grazing animals lead to irrevocable changes. Dense forests become broken into parcels, to be replaced by lighter woodland for coppicing, or by fields, until the forests vanish; all over the world, this process has continued for millennia. Fields were at first small, often irregular, separated by hedgerows or fences, and bearing many mixed crops. In the last couple of centuries, however, and mostly very recently, the development and use of cultivating machines has swept away the hedges, turned small fields into great tracts, many kilometres square, and encouraged vast monocultures. Where animals are grazed in large numbers equally vast extents of grassland have been opened up, as in New Zealand.

Much modern crop cultivation is featureless; among the flat cereal-filled plains of North America, the only prominent features are the circles of bright green crop which indicate the range of huge rotary sprinklers.

The greatest areas of cereal cultivation, notably wheat, are in North America and the former USSR. Only flying above these apparently unbounded fields really brings home their immensity. Every aspect of cultivation, including the addition of fertilizer, is done by machine; weed and pest control is typically carried out by spraying from aircraft. Those who tend these vast crops live in farmsteads or small communities great distances apart, a lonely life in which the understanding of machinery and chemicals remains harnessed to the controlling march of the seasons.

Where fields remain small, and where there is plenty of labour, most aspects of cereal cultivation are carried out by one person behind an animal, or one person on a tractor. In China, wheat is still often cut by sickle, winnowed by throwing the grain up in huge wicker baskets, and finally ripened off on mats in the sun. Although more labour-intensive than mechanized farming, such traditional methods do have the advantage of employing large numbers of people.

Rice, the staple cereal of the tropics, is grown and harvested mechanically in developed countries, but in the steep-terraced fields of the Far East, where the paddies may be only a few feet across as they follow the contour lines, hand labour is the only possible way. In China, work starts by carrying night-soil to the main fields and pouring it in. Then a plough pulled by a water buffalo turns the soil over, both man and beast wading deep in the mud. In separate small paddies where the seed has been densely sown, the workers gather the young plants into bundles. Taken to the cropping paddies, these are placed in position one by one, again by stooped workers with arched backs. Weeding during the growing season is often carried out with the feet; it has been truly said that the traditional farmer knows every inch of his soil. Come harvest, the grain is cut with a sickle, threshed and winnowed and ripened much like wheat. Some warm zones can grow two or three crops a year, so the task here is literally endless.

Tea and coffee

The tea shrub, *Camellia sinensis*, which is a close relation of the garden camellia, is planted in long straight lines over the undulating hills of Assam, Sri Lanka, Indonesia, Japan and China.

Tea is almost always grown as a monoculture but other crops can be interplanted with it, for example cabbages in China; shade trees are also sometimes grown among the bushes. The plant can grow quite large but under cultivation it is kept a little above knee height by pruning and, in particular, regular plucking of the young shoots which pickers cast into baskets slung on their backs; their labour is endless and even more monotonous than that of the rice cultivators. After picking, the leaves go to factory buildings where they are withered, rolled, crushed, fermented (in the case of black tea), sorted, and finally packed. Villages around the plantations are often entirely devoted to the routine of tea, which is continuous in monsoon climates.

Coffee comes from the berries of large shrubs, *Coffea arabica* and *C. canephora*. Picking needs as much precision as tea, because only ripe berries must be taken from clusters which ripen over a period, and the picker has rather more searching to do up and down the plants. After the berries are collected, they may be dried so that their flesh can be rubbed off or pulped entire to remove it mechanically. The beans are then

dried, during which time the final remains of the flesh ferment; this fermentation improves the appearance and also the quality of the final cleaned-up product.

Fruit trees

The widely cultivated apple and pear, grown since ancient times in temperate climates, have been very labour-intensive to cultivate since their early beginnings. These are idiosyncratic trees; although mainly grafted (the selected variety being scions on a stock selected for uniformity and, nowadays, to control ultimate tree size) they develop irregularly, and need annual pruning. But they are grown in rows, which facilitates mechanized weeding and feeding and the application by tractor-hauled blowers of the numerous sprays needed by these insect- and fungus-prone crops each season. Har-vesting is almost entirely by hand, for bruising associated with mechanical picking causes fruit to rot quickly. Large areas are often put down to fruit farms and

a lot of casual labour is brought in at picking time.

The picturesque olive, with its gnarled, often perforated trunks, irregular growth habit and restful greyish colouring is a central feature of many Mediterranean landscapes. Harvesting is done by hand, often by women armed with long sticks, with children to gather the fruits. Nowadays it is usual to cover the ground with plastic sheeting. After harvesting the fruits are taken to oil presses: once powered by circling donkeys, these are now almost all motor-driven. Olive groves are usually maintained by local villagers who often also maintain flocks of sheep or goats. In contrast to the olives, citrus groves often present a regimented appearance. Like apples they are grafted plants, but the rounded trees, planted in endless rows in the Mediterranean and North America, grow more regularly. Again, spraying against the many enemies of this monoculture is essential, and the easily damaged fruit is also picked carefully by hand. Citrus plantations are usually under the control of a handful of

Masses of people join in the final stages of the rice harvest in Bali, as they have throughout the processes of cultivation. Bunches of rice plants, cut by sickle, are banged on the ground to release the seeds; chaff is removed by sieving and winnowing by hand before the grain is bagged up. It is possible to mechanize rice cultivation on level terrain up to a point, but not on the steep-terraced paddy fields prominent in many parts of the Far East. More importantly, the large population, living in villages dotted all over the countryside, needs the work. To introduce automation into such situations is folly.

experts with picking done by casual labour. In America, the migrant pickers move steadily northwards following the ripening of the crops.

Grapes

Wine has been made since very ancient times – perhaps as long ago as 8000 BC; Noah is reputed to have been the first vine-grower. Nowadays there are well over

The symbol of the enormous wheatfields of the northern hemisphere – the world's breadbasket – is the combine harvester operated by a single man. In this land without a vestige of the original vegetation, machines, fertilizers and chemicals against pests, diseases and weeds dominate the giant cultivation operation. Density of population is tiny, for workers live in small communities great distances apart, while the crop goes largely to feed the multitudes packed into cities, mostly divorced from any contact with plants or the land.

10,000 cultivars, from which the astonishing range of different wines is derived.

The grape vine (mainly *Vitis vinifera*) is another very labour-intensive plant, needing expert pruning and training, difficult soil-cleaning operations in often stony ground, and a good deal of chemical spraying. Vine growing is a family affair, or involves small collectives with whole families working over generations. Most attention is by hand, notably picking. Partly because of the plant's end product, harvest is normally a time of celebration, laughter and song (a French saying is that "a mouth busy with song is not busy with the bunch"). In places where the best vintages have been grown for centuries, wine production has stamped the land and its people; little else is grown over large areas. The fields, often terraced and divided by low stone walls, occupy all suitable sites; local variations of soil, situation and aspect all contribute to different end-products from the same grape varieties. In France especially the wine "factory" is often a chateau, itself the centre of social and economic interest. As the grapes mature their progress is the main topic of local conversation, and if disaster strikes – perhaps a heavy rainstorm just before harvest – the neighbourhood is united in despair. Although Europe has traditionally been associated with wine growing, many other regions produce fine wines today, including Australia, California, Chile and Argentina.

Vines also, of course, are grown for dessert fruit in several warm countries and sultanas and raisins are grapes preserved by drying. Currants come from small, mostly seedless grapes, and take their name from Corinth, where the ancient variety black Corinth is still grown.

Ultimately, all human societies depend on plants. The destructive nature of modern agriculture, with its vast monocultures, and prowling combine harvesters, does them much damage. Worse still, the acculturation of primitive peoples, ever accelerating in pace, often takes their irreplaceable knowledge from us for good. The plant world can probably supply answers to some of our most urgent problems – but we have to know where to look. Faced with the task of randomly screening thousands of untried plants, the botanist is like a person fumbling for a key in the dark. The indigenous peoples, however, have hundreds or thousands of years of knowledge to draw on. They hold the key to the plant storehouse.

But though their knowledge is deep, it is not inborn. As communities die out, or are absorbed into the wider society, the oral tradition is broken. When, for example, missionaries in 1959–60 came to aid the Tirios of Suriname, who were being decimated by introduced diseases, the medicine they offered persuaded people to abandon their old way of life. The Tirios have now virtually ceased to use their old cures; the adults recall some details, but the children know practically nothing of the medicinal properties of their own flora. Recorded botanical knowledge is slight, so once the Indians have forgotten, centuries of research will be lost. Similar stories can be told from other regions. For example, the centuries old knowledge of the *amchis*, traditional purveyors of Tibetan medicine in Nepal, is in similar danger of being lost.

Can concerned people act to secure this knowledge? To a degree, we can; and it is greatly in all our interest to do so. Ethnobotany is becoming a priority science, but it is running late in the race against acculturation, and is poorly appreciated. Even if the plant lore of some peoples is saved, the loss of the holistic philosophy of many traditional societies, of their understanding of the natural plant world, is to be mourned. They have much to teach careless western society about the value of plants. According to the South American Indians, legend tells that, "The forest supports the sky. Cut down the trees and disaster will follow."

We should heed them well.

People, plants and fair trade

Global trade and the profits derived from it are unequally distributed in the world. The exports of the 48 least-developed countries, home to ten percent of the global population, amount to just 0.4 percent of total world exports. Contrast this with the exports of the US and Europe combined, with roughly the same population, which account for almost 50 percent. The big multinational companies are also often aided by subsidies, making competition with the small economies even less fair.

The original stimulus for the fair trade movement was the fluctuating price of basic goods such as coffee, tea, cotton, rice and sugar, which had disastrous effects on the local producers. In 2001, for example, coffee prices crashed, with tragic effects on the producers in Ethiopia, a country where coffee is worth nearly 70 percent of export earnings. The story is similar for the banana growers of Dominica, which once numbered 11,000, but are now down to only about 700.

There is a growing awareness in the richer countries of the world that import of goods often means large profits for the companies and outlets involved, but that this is frequently at the expense of a reasonable income for the people who actually produce the goods in their country of origin.

This awareness, promoted in particular by organizations such as Oxfam, has resulted in the gradual development, roughly since the late 1980s, of a fair trade movement. We have also seen the emergence of the ethical shopper, who not only seeks out goods on the basis of quality and price, but who also keeps in mind the welfare of the (often distant) producers. The system works by purchasing direct from farmers at higher, reliable prices, and market ing partly through charity shops and catalogues.

With major supermarkets now also beginning to back such a trend, it is quite possible to buy fair trade products such as bananas, coconuts, tea, coffee, chocolate, marmalade, wine, and oil – even ethically manufactured footballs. Fair trade aims to bypass the middlemen and dealers as far as possible and to pay fair prices direct to the farmers, or to farm co-operatives.

Fair trade thus ensures an equitable and fair partnership between the producers and the consumers. It guarantees that the farmers will receive a living wage, regardless of fluctuations in the global market, and also long-term trade relationships. The money earned is used to invest in health care and education, and gives the local producers economic independence and pride.

To take one example, in Dominica, one of the Windward Islands in the Caribbean, hundreds of farmers are involved in fair trade, which now accounts for over 70 percent of production. Fair trade is often combined with organic farming methods, involving minimal use of chemicals. Fair trade therefore also helps the environment. While the large coffee producers may create extensive plantations (often after felling the forest) and subject their crops to chemical pesticides and fertilizers, the small-scale farmers supported by fair trade agreements tend small plots, often under the shade of forest trees.

Although fair trade is still something of a niche market, like the organic market it is growing, and is now represented in about 50 countries. In the UK for example, sales of fair trade products rose from £92 million in 2003 to £140 million in 2004. The Fairtrade Foundation, which sets standards, estimates that it benefits five million people worldwide. The Fairtrade Federation, based in the US, is an association of fair trade wholesalers, retailers and producers, whose members are committed to provide fair wages to farmers and other producers worldwide. In the US, for example, there are now over 100 companies that offer fair trade certified coffee, with more than 7,000 outlets, and this is growing. The various fair trade organizations together run Fairtrade Labelling Organizations International (FLO), based in Bonn, Germany, which sets international standards and monitors the trade. By September 2004 there were 422 certified producer groups in nearly 50 countries, selling fairly traded products to hundreds of importers and retailers.

Benefit-sharing

Related to the fair trade problem is the more general issue of the need for the equitable sharing of any rewards arising from the development of products, recognizing the rights of farmers and other local people. This applies particularly to medicinal and agricultural products exploited by companies, and often based upon local, traditional knowledge.

In June 2004, the International Treaty on Plant Genetic Resources for Food and Agriculture came into force. This recognizes the contributions of farmers, and identifies ways of protecting their rights, and ensuring that the benefits are shared with the countries in which they originated. Making sure that this actually happens in practice is difficult, but at least now there is a general acceptance that this should be the way forward, and that the local holders of specialist knowledge should not lose out when their expertise is exploited and commercialized.

The pharmaceutical industry has been one of the major exploiters of traditional knowledge. Nearly three-quarters of plant-based drugs were discovered as a result of ethnobotanical research, and are thus based upon local expertise. The picture is further complicated by the use of patents by companies, claiming ownership of the properties of organisms, including in some cases medicinal plants, that have been central to local communities, often for centuries or even thousands of years.

There are however, success stories. One example is the agreement reached involving the government and local people in Samoa in the case of the tree *Homalanthus nutans* (Euphorbiaceae). This plant has been shown to contain prostatin, a promising anti-HIV compound, the bark of which is used locally to treat hepatitis. The Aids Research Alliance of America agreed to return to the people of Samoa 20 percent of commercial revenues received from the use of prostatin.

Guidelines are now available for companies wishing to work with local experts, and pressure towards equitable benefit-sharing, from governments and advisory bodies, is increasing. In future it is to be hoped that the use of local, traditional knowledge by distant commercial enterprises will increasingly be on the basis of a shared, mutually beneficial approach.

Cocoa production in Nicaragua. A cocoa worker inspects his crops in the town of Waslala, some 208 miles north of the capital Managua. The production of cocoa has become an alternative for the subsistence farmers who prefer its profitability to the fluctuations of international coffee prices. Small farmers currently export some 200 tonnes of organic cocoa to Europe per year.

CHAPTER TEN

Improving the Resources

Thousands of people die every day of starvation or diseases related to malnutrition, and some have predicted that by the year 2025 food production will need to have doubled to keep pace with the increased demand.

Few people realize how vital wild plants and primitive crop varieties or "landraces" are to plant breeders. Yet our hope for the crowded world of the future rests on just such plants, and the untried virtues they may offer of disease-resistance, drought or salt tolerance, ease of harvesting, or soil improvement. We can only guess how many of these potential saviours are in danger; the majority grow in countries of rapidly rising populations and extreme poverty, where the pressure on them increases year by year.

One way to save these plants is by in-situ, "on-the-spot", conservation – protected areas like nature reserves and national parks. Another is ex-situ, either "in cultivation" or "in-the-refrigerator", conservation – gene banks for seeds that will remain viable in cold storage, especially important for maintaining the seeds of cultivated varieties.

As long as we can protect these resources, there is hope that the partnership of modern science and the plant kingdom can flourish, and continue to provide all we need. New skills exist in breeding and manipulating plants to suit uncongenial environments or to supply specific food or industrial requirements. Laboratory techniques, too, offer the prospect of breeding large numbers of plants at once, or of persuading them to mass produce their valuable chemicals for us.

There even is the possibility, in some measure at least, of increasing food produc-

Domestication and selective breeding have produced many varieties of tomato, in a range of colours, flavours and shapes.

tion and supply of industrial substances, without putting pressure on land, by turning to products made by micro-organisms. Industry is now cultivating bacteria to provide an animal feedstuff in massive fermentation plants. As foods, micro-organisms are not as unfamiliar as they may sound. Yeasts and other fungi have been involved in the production of bread, cheese, yoghurt, beer, wine and vinegar since ancient times. In Indonesia, fermented soybean and peanut cakes make up a third of the protein requirement; the similar bean curd is widely eaten in China. The use of the cyanobacterium *Spirulina* as food is described on pages 78–9.

Bio-engineering is a high-tech solution which can, at best, only serve to reduce the pressure of hungry populations on our remaining untouched natural vegetation. Without the wild plants and ancient landraces, we have no wealth for the future. Conserving these, in one way or another, is the first imperative.

The age of the bug

Modern biotechnology, the *industrial* processing of materials by micro-organisms, has achieved much in its short history – new foods and nutritional upgrading of traditional foods; antibiotics, vaccines and drugs for human and animal use; enzyme technology for industrial and clinical purposes; energy and fuel production; pollution management; extraction of metal from waste and low grade ores; and polymers for use in food, textile and heavy industries.

Crops produced by biotechnology are now grown in every continent, with the global commercial value of such crops estimated at US$ 44 billion in 2003–04, with 98 percent of that value produced in five countries – United States, Argentina, China, Canada and Brazil. The main crops involved are soybean, cotton, corn (maize) and canola (*Brassica napus*, for oil). In the same period the value from such crops in the US was $27.5 billion. In the next ten years, the global value of biotech crops is expected to reach over $200 billion. China is now a major centre for biotechnological research. But in Europe, approval of biotech crops has been restricted, and there is widespread suspicion about genetically modified crops and their possible effects on the environment and health (see also Genetic modification – solution or problem?, p. 171).

Test-tube plants

Many plant substances highly prized in medicine or industry require long and complex extraction processes. Some day, perhaps, these may be manufactured in test tubes, by plant cells. If a piece of plant is disinfected, broken down to single cells or small groups of cells, and grown in a carefully composed nutrient medium, a mass of undifferentiated tissue will form (in the right conditions). With luck, these cells will release the desired substance straight into the culture medium.

The possibilities are enormous. The pharmaceutical industry, for instance, derives a quarter of all prescribed medicines in the western world ultimately from plants – more in the east. The rosy periwinkle (pp. 118, 120), used against cancer, has already been grown in a test tube: the cells synthesize the valuable alkaloids, scarce in the plants, at high rates and secrete them directly into the culture material, greatly simplifying drug extraction. Contraceptive pills could also be manufactured this way, using cell cultures of the hormone-producing yam species (pp. 121, 122) – particularly desirable in view of their increasing rarity in the wild.

Many factors induce drug firms to invest in cell culture: unpredictable harvests at limited times of the year; unpredictable supply in changing political situations; the cost of importing plant materials from the other side of the world. But the science is still in its infancy. In western medicine, compounds from plants are either incorporated into drugs directly, or the active chemicals are used as blueprints for synthesis of related compounds which may even have a greater effect than the original. At least 30 commonly used drugs are derived directly from plants, and the global sales value of plant-derived medicines stands at around $30 billion. About 80 percent of the world's population relies wholly or partly on locally produced medicines, mostly of plant origin.

Tissue culture has other exciting applications, too. If the cells (from buds, leaf segments or roots for example) multiplying in culture are encouraged to reorganize, it is theoretically possible to produce plantlets – and so breed up huge numbers of plants. This process, known as micropropagation, produces cloned plants that are genetically identical to the parent tissue. The technique could help us reafforest huge areas, regreen denuded lands, take pressure off exploited wild plants, and increase food supplies.

All the cereals, however, belong to the monocotyledons, which are notoriously awkward customers when it comes to tissue culture, and do not regenerate at all readily. It is only recently that a breakthrough has come: scientists are now regenerating palms (also in this group, and normally only grown from seed) in culture. Cloning a high-yielding individual oil palm in this way may increase yields by 20 to 30 percent; yields from cloned coconut could be increased fivefold.

One special kind of tissue culture is meristem propagation, in which almost microscopic shoot tips, like miniature cuttings, are placed in flasks of growing medium and constantly turned to remove gravitational effects. Each tip rapidly produces clusters of tiny plants, which are later divided and grown individually. The method has been used successfully with orchids, enabling growers to build up stocks of new, expensive hybrids quickly.

Breeding for improvement

Until recently, the only means of improving plants has been to select those with useful natural mutations – a process which has been going on in farmers' fields for millennia – and then deliberately hybridize (cross-breed) the best cultivars which emerge. But today, the fledgling science of genetic engineering offers the prospect of inserting foreign gene material directly into a new host, thus modifying its makeup – genetic modification (GM). Scientists are working on techniques that will cross the barriers of sexual reproduction, even strad-

Scientists now use laboratory techniques to save endangered plants. Here, seedlings of the rare Australian orchid Pterostylis fischii *are ready to be planted out, having been raised in test tubes. The orchid seeds, produced in abundance, are tiny and only germinate in the presence of a fungus which provides essential nutrients. The scientists' method involves growing the seeds in a weak porridge inoculated with the fungus. Six endangered Australian orchids have been bred by these techniques, and reintroduced to nature reserves after their brush with extinction.*

dling the divide between the different kingdoms of plant, animal and micro-organism. Inducing cereal crops to take on the nitrogen-fixing properties of the legume – *Rhizobium* symbiosis (pp. 101, 102) is one, at present distant, possibility. But though full of unexplored possibilities, which will certainly surface in the years ahead, genetic engineers can only supplement the standard methods of improving plants: we cannot create genes, and so still need to use existing plants to derive new strains. While thought by many to offer great potential benefits, GM techniques have been fiercely attacked because of perceived potential dangers to the environment and to human health. These issues are discussed in more depth on pages 171–3.

There are two important considerations in breeding. The first is obvious: in crossing two plants, each with distinct desirable characters, the breeder hopes to combine the virtues of both in their offspring. Sometimes the attempt is to breed just one valuable character (a specific disease-resistance being the most obvious example) into an existing cultivar which is otherwise excellent. The second purpose of breeding is to create an F_1 (first generation) hybrid: if two selected cultivars are crossed to produce seed, the plants grown from this seed often (though not necessarily) have increased vigour (often termed "hybrid vigour"), displayed in improved stamina and increased yield. The catch is that the F_1 hybrid cannot be reproduced from its own seed; the second generation would show very great variation. The cross to produce F_1 seed has to be repeated annually (it usually applies only to annuals –

garden flowers and most cereals). This places the propagation of an F_1 strain in the hands of its creators – the farmer must buy new seed each year.

Plant breeding and production of new strains have been vastly stepped up recently partly because computers can now monitor enormous numbers of crosses, and a wide range of characters.

Genes from the wild

All cereal crops have in the past acquired improvements through accidental breeding with wild relations, and most have been deliberately improved, too, with genes from the wild. Even so, this resource remains partly untapped. Breeders will resort to the wild only if the crop to be improved is rather uniform and therefore offers insufficient genetic wealth (known as variation) for selection, or if the desired qualities are much more pronounced in the wild forms than the cultivated. It can be very difficult to make a fertile cross using wild material; and even if a cross is achieved, it may not be easy to get rid of the weedy characters associated with the desired quality. Besides which, the culti-

vated forms are so much better known and more readily available. Nevertheless, the genetic richness of wild crop relatives remains a source of great potential value and such species are worthy of special efforts to conserve them.

The tomato *(Lycopersicon esculentum)*, owes its enormous commercial success almost entirely to its wild relatives. With a little help from its friends, the crop has gained: resistance to the fungus *Fusarium oxysporum* (borrowed from *L. pimpinellifolium*); a change to the fruit stalk that adapts it for mechanical harvesting (from *L. cheesmanii*); intensified colour and increased soluble solids (from *L. chmielewskii*, among other species, even though *chmielewskii* is green); higher betacarotene (*L. hirsutum*) and increased vitamin C content (*L. peruvianum*). The tomato is a rather unusual crop in that it apparently was first domesticated not in its ancestral home in Ecuador, but in Mexico where it had travelled as a weed in maize crops. This alien origin deprived its first cultivators of access to the wealth of genetic variety normally available to a species.

The genetic wealth, or gene pool, of a crop comprises all the genes from all the varieties and species that could possibly breed with it. The *primary* gene pool consists of domesticated varieties, the wild original, and related wild species that breed readily with the crop to produce fertile offspring. The *secondary* gene pool includes the species that will cross with the crop using conventional methods to give some fertile, but many sterile, progeny. The *tertiary* gene pool requires the special techniques of modern science to persuade the reluctant parents to breed; even then the hybrids will usually be completely sterile. Tomato and cotton at present are unusual in having had genes added from all three pools.

The wild genes most widely used are those for disease-resistance, usually major genes which confer vertical resistance – that is, protection against a specific race of pathogen. The wild potato species *Solanum demissum*, for example, from the secondary gene pool, was used originally for its major gene resistance to late blight. However in 1936, races of the disease arose against which the gene proved ineffective, so breeding with *S. demissum* was largely abandoned. But then it was noticed that varieties with *S. demissum* in their background were generally a little more blight-resistant than those without. Generalized or horizontal resistance of this kind is conferred by many minor genes, and is now considered an important safety net to prevent new races of disease wiping out crops.

Genes from the wild have been used in all of the big three cereals. Bread wheat has already obtained considerable benefit, including resistance to eyespot disease *(Cercosporella herpotrichoides)* from goat-grass, *Aegilops ventricosa*, and frost resistance from *Agropyron*, while commercial maize has been improved with genes from the wild *Tripsacum dactyloides* (p. 71).

The wild input into rice has also produced improvements, sometimes with spectacular results. The big new paddies of the "Green Revolution", cropping rice all year round, greatly suited the ecology of the grassy stunt virus, which never found itself homeless, as it might if there had been a break between crops; the rice was also an abundant food source for the virus's carrier, the brown plant-hopper *(Nilaparvata lugens)*.

Grassy stunt appeared almost from nowhere as an important disease. In the early 1970s, 116,000 hectares of rice in Indonesia, India, Sri Lanka, Vietnam and the Philippines were destroyed by epidemics of this virus. Immediately, the hunt was on for a high-resistance gene. Screening of the 6,273 samples of rice kept in the collection of the International Rice Research Institute revealed that only one, the feeble *Oryza nivara*, had the required character; as a bonus, it was also resistant to blast, another major disease of rice. This sample had been collected from central India in 1966, at a time when no one could have guessed that it would ever be of value – grassy stunt was of no importance then and the wild rice had lots of bad points – weak stems, low yields, with a head that shattered at maturity to scatter the seeds. Now, thanks to *Oryza nivara* (fortunately in the primary gene pool that breeds readily with the crop) the disease has gone back into obscurity, and IR36, the rice variety with the wild genes, is planted across millions of hectares. Although *O. nivara* has vanished from its original discovery site, its genes persist in many modern rice varieties. Advantages of this kind can be of temporary benefit, and pathogens evolve new strains, resulting in an arms race, requiring scientists to breed new varieties to stay ahead of the game. Another wild rice, *O. rufipogon*, found on acid soils in Vietnam, has been used to create a new variety of commercial rice that is now grown on soils too acid for most other types. One famous variety, dubbed "golden rice", has been genetically engineered to produce betacarotene, with the aim partly of combating diseases such as blindness caused by vitamin A deficiency. It may also be possible to produce a non-GM vitamin A rice by identifying and using natural genes for betacarotene production in rice. The first accurate map of the rice genome has recently been produced by Chinese scientists, and this should greatly help the crop breeders.

Rice breeding has benefited thousands of farmers, and desirable traits such as better grain quality and faster ripening (allowing a second crop such as a legume) are a feature of some new strains.

Cocoa provides another intriguing story. Most is produced in Africa from old plantations of unimproved stock; but since demand for the commodity is still rising, the trend is to develop new, high-yielding hybrids. However, cocoa cultivation spread around the world so quickly that all the crops were originally derived from just a few wild ancestors, and so suffered from a lack of genetic variation. Faced with this, commerce had to go back to nature: wild and semi-wild material has revolutionized cocoa breeding, providing vigour, resistance to the crop's main diseases, and greatly increased productivity. The wild species (*Theobroma cacao*) grows as trees or shrubs in the diminishing tropical forests of the Amazon, with greatest diversity along the eastern fringe of the Andes, as in Ecuador. Mature trees can reach 20 metres, but most crops are pruned to about five metres. Human use of cocoa goes back a long way – it was considered a sacred plant by the Maya and predecessors and also the Aztecs. Many different varieties and cultivars now exist. West Africa produces the most cocoa, with Ivory Coast the biggest producer at over 100 million tonnes per year. Other main producers are Ghana, Indonesia, Nigeria and Brazil.

Many of our major world crops have been improved with wild germplasm: these include rice, barley, oats, sweet potato, sunflower, oil palm, sesame, tomato, carrot, bell or sweet pepper, grapes, apple, strawberry, sugar cane, sugar beet, rubber, cocoa, cotton, maize, potato, cassava, peas, tobacco and wheat. The use of wild species in sunflower breeding has helped to raise yields by 20 percent, and in sugar cane breeding has almost doubled the cane yield and more than doubled the yield of sugar. Some crops, such as the cabbage tribe, already have so much variation among the cultivated types that it may not be necessary to resort to the wild. Yams are similar, but wild species have much potential. A number of minor crops, too, have benefited from their uncultivated relatives, including hops and gooseberries.

Landrace genes

Even more important to the plant breeder are the landraces – the traditional varieties of crops that evolved under human use over the centuries. Developed in an environment with relatively low levels of cultivation, fertilization and plant protection, these varieties experienced selection pressures for hardiness and dependability, rather than productivity. Moreover, the genetic diversity found naturally within each landrace provided the farmer with insurance against climatic extremes and disease epidemics. Traditional varieties, in fact, are often able to withstand conditions that would seriously harm many modern lines; but they may be so closely adapted to their particular locality that they do not do well in other environments.

Within these landraces lies a wealth of genes, just waiting to be sorted into the right packages; so the first purposeful selections led to greatly improved yields. In the minor crops, especially in the tropics, there is even now considerable scope for selection directly from landraces. But the greatest benefit they offer today comes from the individual genes they contain, which can be inserted singly into ailing crops to revitalize them. Landraces have been described as the product of human genius, and their survival has depended upon the caring hand of rational agriculturalists and traditional farmers down through the generations.

Naturally enough, the centres of genetic diversity of crops and of their pathogens (disease organisms) often coincide – the Old World cereals, and the rusts, are an example. The diseases develop to exploit the plants, and the plants evolve countermeasures to combat the diseases; but they manage to rub along together. It is not in the interest of the disease to exterminate its host; the two co-evolve. Disease resistance can also develop in landraces far from the crop's centre of origin. A survey among 6,689 samples of barley in the US Department of Agriculture's world collection revealed only 117 with resistance to yellow dwarf virus – 113 from Ethiopia, three with Ethiopian parents, and one from China; but the crop originated in the Near East. It can safely be assumed that this rare asset arose in Ethiopia, not in the crop's original home.

In 1974, a group of scientists went to the mountainous areas of northern Moravia and Slovakia. They gathered material from fields, gardens and farmers' stores and found 247 different cereals, grain legumes, vegetables and medicinal plants. The expedition was in the nick of time; the remaining material is vanishing fast. The old landraces are disappearing everywhere, overtaken, and ousted, by the introduction of "improved" and F_1 strains.

The new high-yielding varieties of wheat and rice are causing special concern in this respect. As the "Green Revolution" varieties of wheat and rice spread around the world, farmers eat the seeds of their old landraces, and plant the new. As geneticist Garrison Wilkes has said, this is like "building the roof with stones from the foundation". In Pakistan, all the areas to which access is easy, by road or path, have reportedly been planted up with the new cultivars. Even in the Himalayan valleys, isolated by high mountain ranges, the ancient races of wheat

The soybean (Glycine max) *depends on a slim genetic base. In the USA most varieties are derived from just 11 lines. Hitherto the increase in production has been brought about largely through increasing acreages; but with only a finite amount of land available for the crop, agriculturalists now wish to raise the amount produced with genetic improvements. The worry is that the variation does not exist within the crop, and that the oriental germplasm base needed for the improvements is fast vanishing. The homelands of the soybean's wild relatives are being covered by building and industrial developments and landraces are being displaced by higher-yielding selections.*

are being replaced with high-yielding lines from the south. In Nepal, the wheat-growing area has increased dramatically, with 80 percent given over to the new varieties. Collectors must push ever further on to remote places to find old races, and breeders will have to work urgently to collect the surviving range of types before they are exterminated. In the last 50 years, 95 percent of native wheat varieties have been eliminated in Greece. There has been equally catastrophic genetic erosion in Afghanistan, India, Iraq and Turkey; and this great loss has all occurred in the last 30 years. The case of rice is also alarming. Rice feeds three billion people, and by 2025 this will rise to 4.6 billion. Over the thousands of years that rice has been cultivated, thousands of local varieties have evolved – India alone had about 200,000, but these have been seriously diminished and replaced by commercially produced, "improved" varieties. At least 120,000 varieties are now recognized.

So the "Green Revolution" has been a mixed blessing, some arguing that it has led to excessive use of agrochemicals and the need for high fertilizer inputs; and this is in addition to the fact that in many areas the new plant varieties have displaced local landraces, thus reducing diversity. One of the challenges for the future will be to take advantage of the new technologies and scientific developments in crop breeding without losing the genetic richness of local, traditional crops.

The threat to wild crop relatives is equally alarming. Take maize, for example: the cleared forests and verges of new highways,

opening up across Mexico and Central America, may sometimes provide a temporary habitat for its weedy ancestors the *Tripsacum* species and teosintes (notably *Zea perennis* and *Z. diploperennis*); but in the course of construction, the ancient stands of these wild grasses may be destroyed, and human settlements following the roads continue the attrition. The up-to-date farmer uses pure, hybrid seed in large, weed-free fields, destroying all the old varieties and squeezing out the weedy erstwhile bedfellows on the old field margins. In some places in Central America, maize land has been cleared and planted with more lucrative cash crops, notably strawberries, for export to the United States.

Genetic loss is occurring in many other crops. Recent reports have pinpointed wild relations of cocoa, coffee, sugar cane, sweet potato, tomato, some pulses, pepper, onions and brassicas as all seriously endangered. This wild germplasm is the fundamental resource of biological engineering – and in real danger of being the limiting one. No one can (as yet) create genes with specific attributes.

The world over, wild relatives of crops are being decimated, from cabbages in France and Italy to cotton in the US, Ecuador and Peru. The oats of Morocco and Spain are under threat, sunflowers in the US, avocado in Ecuador, olives in North Africa, oil palm in South America, beets in Portugal, Greece and especially Turkey, safflower in Morocco, and coffee in Ethiopia. In western Asia, the wild pears, cherries, pistachios and other fruit and nut trees are being felled for timber and fuel. In

South-east Asia, the lowland tropical rainforest, home of the wild relatives of mango (Mangifera), durian (Durio) and rambutan (Nephelium), is being cleared.

The current loss of genetic resources, both landraces and wild crop relatives, can be attributed to three main causes. The chief of these is loss of habitat, the second is over-exploitation, and the third competition or predation by introduced species. Of the 250,000-plus (and possibly over 420,000) flowering plant species, as many as 75,000 may be threatened with extinction. Who knows how many of these threatened species are potentially useful to us? This unknown quantity of useful plants is in danger of genetic erosion before domestication has even begun. And there are also species that have only just entered our catalogue of value, and as yet differ little from the wild, such as the Monterey pine. There are only a few small populations remaining in its native California, yet it is an important plantation tree in Australia and New Zealand. The remaining Californian specimens are clearly a great genetic treasure. The wild Douglas fir of the Pacific coast is threatened by plantations of the domesticated form which, shedding their pollen into the common pool, could reduce the diversity of the indigenous population.

About 75 percent of the genetic diversity of agricultural crops has been lost during the last century. For example, more than 7,000 varieties of apple were grown in the US in the 1800s, yet all but about 300 of these are now extinct; now only about 1,000 varieties of wheat are grown in China, whereas formerly this stood at 10,000 types.

Conserving the resource: in situ

Wild crop relatives can be conserved in situ by setting aside the areas where they live as reserves, with facilities for study and collection of germplasm under licence. This is a relatively recent use of reserves – most protected areas in the past have been set up for their scenic beauty, to protect individual species, or to maintain representative ecosystems. An important part of current conservation efforts is to develop management methods and goals for in-situ gene conservation.

Not surprisingly, very few areas so far have been set aside for this sole purpose. There is one in the Kopet-Dag mountains of Turkmenistan, to protect fodder grasses, apricot, pistachio and almond, and one in the Caucasus for wheat and fruit trees. Sri Lanka has a reserve for medicinal plants and India is planning reserves for banana, citrus, rice, sugar cane and mango. Most reserves also serve other purposes – a good thing since the price of protection can be spread across other objectives, such as amenity and watershed protection. It is helpful to the breeder to study the ecology of the plant in its proper habitat. But the greatest strength of in-situ conservation is that the crop relatives continue to evolve alongside their pests and pathogens, under the pressures of natural selection. They thus provide the breeder with a dynamic source of genes.

In-situ reserves are ideal, too, for plants whose germplasm is difficult to store artificially. Some seeds cannot be dried, or have brief viability; some plants seldom produce seeds; some seeds rarely germinate. These genes have to be stored as roots, tubers, cuttings, or whole plants, which occupy too much space for a good range of genotypes to be held. They are more easily conserved in the wild.

Organizations such as the International Plant Genetic Resources Institute (IPGRI) are working to encourage the conservation of wild relatives of crops, including progenitors of crops and species more or less closely related. Together these plants are an important resource for the future improvement of sustainable agriculture. Modern varieties of most crops now contain genes from their wild relatives.

In fact, very few protected areas have been established specifically to conserve plant diversity, and such reserves tend to be found in countries where there is interest in rare or endangered species. Small protected areas ('micro-reserves') are being increasingly established for plant conservation, notably where human pressure on the land is high. Some worry that the small sizes of populations of rare species in these reserves makes them vulnerable to extinction. But some of these species have probably always been rare, and may be adapted to low population sizes. In the Mediterranean region, many species have very restricted ranges, being confined just to one or a few sites, such as mountain tops, cliffs and dunes. There is evidence that even small fragments of tropical forest are valuable for conserving plant diversity, albeit a different set of species than found in larger forests.

One example is the establishment of micro-reserves in the Valencia region of

Spain. Here no fewer than 97 percent of the 350 species that are endemic to the region do not grow in its widespread vegetation types, such as garrigue and pine forest; rather, they are found in specialized, often small (1–2 hectare) sites, and often isolated from one another. By 2004, 230 such reserves had been created, covering a total of 1,440 hectares and including more than 70 percent of the endemic species. These micro-reserves have been established on both public and private land, always with the agreement of the landowners. They are voluntary, but the agreements, once made, are irrevocable and the reserves cannot be deregistered by landowners. A one-off payment is made to landowners to compensate them for loss of economic activities; funding is also available for conservation tasks. Another example comes from Tanzania, where the government has announced plans for a new national park of 135 sq. km on the Kitulo Plateau in the Southern Highlands. This is a grassland area with a rich flora; if realized, it will be the first national park in tropical Africa established specifically to protect plant diversity.

Conserving the resource: ex situ

The loss of genetic variety that has taken place in the last few decades has underlined the need for places where plant genes can be stored for safe keeping and future use – "gene banks". The first great collection of germplasm was assembled by the Russian botanist N. I. Vavilov and his colleagues in the 1920s and 30s. He and his successors set

a great example that has been emulated the world over, but not always with equal success or scientific rectitude.

The collections are usually made up of seeds, and they need to be kept in cold storage at very low humidity, if they are to keep well. Unfortunately, this introduces a new and important selection factor – the ability to withstand the very conditions of storage. Thus, particularly at the end of an unusually long period, the remaining viable seed will probably be adapted to storage, and not be a fair reflection of the original sample.

The seeds of most species will keep in cold storage for a number of years, but they are living organisms, and must be tested for viability regularly and also sown and grown (known as growing out) once in a while or they will die. Self-pollinated crops can be grown out in garden plots, but open-pollinated species need to be raised in bee-proof glasshouses and hand-pollinated to ensure that the lines remain unadulterated. Even some very famous collections lack the facilities to plant out as regularly as they should, and it is rumoured that some seed banks now contain large amounts of rotting or dead material.

Other, more obvious, troubles are caused by pathogens attacking both seeds and vegetative material. Losses occur too from causes like refrigerator breakdown, electricity failure, transferring material in unsuitable conditions, or growing out in the wrong way (one maize strain has been killed off by being grown at the wrong altitude). It has been calculated that only 11 percent of gene bank storage is truly long-term.

Seed material is generally preferred because it occupies little volume. But some seeds, notably those of certain tropical plants, have no dormant period and will not keep beyond a few weeks, so germplasm from these must be stored as vegetative materials (roots, cuttings, and so on). Cassava and yams, for instance, are banked this way; so are bananas, and some peanuts, both of which practically never produce seed. Most tropical fruits and most medicinal plants are maintained by growing specimens from cuttings. Vegetatively produced materials are no problem as far as genetic purity goes, since they are produced without the genetic shuffling of sexual reproduction, and are always identical to the plant from which they were taken. Even so it appears that much variability can be lost between each growing-out.

Many working collections kept by plant breeders have only the small range of varieties likely to be easily assimilated into the breeding programme. Sometimes the assemblage lasts only as long as the different lines are being investigated, and the collection is abandoned once the final selections have been made; it takes time, space and money to maintain a good collection. At the other end of the spectrum are conservation collections such as those held at the National Seed Storage Laboratory in the USA and the N. I. Vavilov Institute in St Petersburg, Russia, which cover a substantial part of the genetic variation of a wide range of species. The Millennium Seed Bank Project at the Royal Botanic Gardens, Kew, aims to safeguard 24,000 species from around the globe, by 2010. IPGRI maintains databases that hold information on ex-situ collections worldwide, and this currently stands at five million accessions representing more than 20,000 species.

There are now national gene banks in many countries, as well as private seed companies, and specialist crop research centres. The International Potato Centre in Peru, for example, was set up to find a mutation that would enable potatoes to flourish in tropical lowlands. Based on the dry coastal plain, it runs several other stations where the wild potatoes are grown, both in their natural habitats and in experimental ones, such as tropical forest.

Rice also has its own bank. The International Rice Research Institute in Manila, Philippines, now holds 90,000 samples of cultivated rice and wild species from more than 110 countries in its gene bank, as insurance for the future. Its aim is to "preserve for posterity the fruits of thousands of years of natural and human selection".

The centre for world maize and wheat collection, with 120,000 entries from 47 countries, is the International Maize and Wheat Improvement Centre in Mexico, and research is also carried out at the Indian Agricultural Research Institute in New Delhi. The Asian Vegetable Research and Development Centre (World Vegetable Center) on Taiwan conserves the most diverse collection of vegetable germplasm in the world. It has more than 50,000 accessions of 334 different species from over 150 countries. These include mung beans, common bean, Chinese cabbage, cauliflower, mustard greens, radish, pepper, tomato, and sweet potato. The National Vegetable Research Centre at Wellesbourne, England, has a comprehensive vegetable gene bank. It holds seeds of cabbage, sprouts, calabrese, cauliflower, carrots, celery, beetroot, turnips, swedes, parsnips, onions, leeks, rhubarb, tomatoes, lettuce and smaller quantities of other vegetables, including those from the tropics.

Several British stately homes have become vegetable sanctuaries, and started growing "heritage" seed. In the USA, old vegetables discontinued by the commercial firms are handled by small "heirloom" firms, trading maizes, for instance, that rejoice in such names as bloody butcher, Gila Indian, Aunt Mary's sweet, extra early Adams and pencil cob. Enthusiasts can join clubs, such as the Seed Savers Exchange, to swap their heirloom varieties.

Some gene banks are in commercial hands, and so subject to vagaries of profit and loss, while others are poorly or unsuitably housed. Good gene banks cost a lot; most of the money and effort in gene conservation goes into ex-situ, not in-situ, conservation. But the international community surely needs to upgrade them, establish new ones, and safeguard their resources. A key initiative is the Global Crop Diversity Trust, launched in 2002. This undertakes to raise funds to set up gene banks for the world's most important crop collections. There are already six million crop samples in 1,500 gene banks around the world.

Who owns the genes?

Quite suddenly, it seems, those who control and cultivate world food crops are realizing that all is not as it should be. In attempts to improve existing crop varieties, breeders seek landraces and weed relations which contain the biggest potential, or often the biggest safety net – only to find that the new seeds of the past few years, the modern ways of cultivation, have removed immense numbers of them from the earth forever.

Even so, many old strains remain with us. Some are now lodged in gene banks or in reserves; some still exist in the wild, and should be sought out. But just who do the genes belong to? For example, in the 1960s, wheat crops in North America were attacked by the deadly disease stripe rust. Breeding from wild wheat found in Turkey produced new strains resistant to the rust, so that yields were not just maintained, but increased. The US Department of Agriculture has estimated its resulting annual saving at $3 million. But Turkey has not been paid for its wild wheat genes. Or has it? Since 1970, when the improved wheats were made available to Turkey, their cultivation there has doubled yields; farmers have received more cash, the Turkish government has presumably benefited from profit taxes, and more people have been fed.

Other examples include wild cherry material, collected in northern China, which has virtually resuscitated the cherry industry in Britain, and wild lucerne material, collected in Libya, that one Australian breeder has said is worth millions to the country's livestock industry.

The 1992 UN Convention on Biological Diversity (CBD) aimed to encourage the free exchange of scientific information and dissemination of genetic material, as well as stressing the need for equitable sharing of financial rewards resulting from the development of products. This applies as much to agricultural as medicinal products. In practice, equitable sharing has not always been achieved but there is general agreement that this should be the goal.

Some countries are already creating administrative obstacles to foreign botanical collecting teams. Mexico has actually banned the collection of most wild plants, and Ethiopia has legislation to stop coffee germplasm leaving the country. At least 55 percent of all developing-world crop genetic resources in storage are housed by the developing nations. The germplasm has been sent on the assumption that it will be freely exchanged; but this is not necessarily the case, and many countries limit its flow.

Large companies are buying up seed firms so that they may each control great numbers of breeding and distribution arrangements. A general increase of corporate control of agriculture has been a feature of recent years, and one aspect of this is the attempt by biotechnology companies to patent plant products and food crops. For example, 250 patents, mostly corporate owned, had been granted on varieties of rice by the end of 2001. Such commercialization of plants and their products often runs counter to the encouragement of local, traditional knowledge, experience and expertise.

Less profitable varieties in the seed lists of the seed firms bought up can be lost for ever, despite their genetic potential. These varieties are often those best suited to limited areas of cultivation.

The involvement of large seed companies in pesticide and herbicide manufacture has led to concern that they might stop trying

to incorporate genes for resistance against diseases, eelworms and so on into crops, and then direct customers to their own-brand pesticides when disease threatens. They might furthermore design seeds so that only their own-brand chemicals are effective: others could even be toxic to the crop. At the very least, they might offer the farmers packages of seeds and chemicals, thus encouraging heavy use of pesticides. Credit, and cultivation facilities such as irrigation, may also be linked to the use of specific seeds, when offered to the farmers by developing-world governments who deal with these companies.

One result of such package deals is the coated seed. A disturbing example is that of grain sorghum marketed by one company. This is coated in three chemicals, two of which protect against grass invasion and the third against the company's own herbicide, which is toxic to sorghum. Apart from such extraordinary situations, there is always the danger that in times of famine, seed grain gets eaten – and if it is coated those who eat it will join the 10,000 peasants who die from pesticides each year, of the 375,000 who become ill from their misuse.

Another famous example concerns the popular weedkiller "Roundup" (glyphosate). This chemical is used all over the world to kill "weeds" – on farms, in gardens and also for example on golf courses. Since the late 1990s crops have been created which are so-called "Roundup-ready", that is they have been genetically altered so that they are not killed by Roundup. Using such crops (marketed by the same company that markets Roundup) farmers can spray these

herbicides without harming the crop. The main crops affected are soybean and cotton. But, aside from some reported health problems caused by exposure to glyphosate, such liberal and systematic use of the herbicide is now creating herbicide-resistant weeds. Thus the introduction of Roundup-ready crops may be rendering the herbicide itself less and less useful. This situation is analogous to that of antibiotic resistance in medicine.

Plant Breeders' Rights (PBR), or plant patents, have been in force since the 1930s, and have now been established in some 25 countries. They allow plant breeders to obtain royalties, either from seeds they have produced or from plants themselves. It seems reasonable that years of careful research and breeding should be rewarded, but the potential damage against the genetic resource is unfortunately considerable. Minor varieties are ignored, because PBR demands a unique factor in the new cultivar. This also works against the production of "multi-lines", strains that are uniform and stable but have enough intrinsic variability in matters like disease-resistance to be preferable to a single strain. One cannot patent a multi-line, only its components, which is very expensive and seldom done.

Further, many grow-outs needed regularly by gene banks are handled by commercial firms. In this way, unfamiliar genes are in effect offered to commercial breeders to assess. Seed from grow-outs, and also from international agricultural research centres, has undoubtedly been purloined, grown on and patented by seed firms, with no credit or payment.

PBR is difficult to police. Some traditional cultivars and some landraces have certainly been patented in developed countries and, as they stand, the PBR laws encourage such poaching. Even entire botanical species can be "protected" in this way so that, for instance, a deep-coloured heliotrope has, after selection from the offspring of collected seed, been granted PBR with the result that Guatemala now cannot export this quite common wild flower to the US.

In 2002, the Convention on Biological Diversity (CBD) agreed to a Global Strategy for Plant Conservation, setting targets to be met by 2010. These targets include the conservation of 70 percent of the genetic diversity of crops and other plants of major socio-economic value, and the maintenance of associated local and indigenous knowledge.

In June 2004, the International Treaty on Plant Genetic Resources for Food and Agriculture came into force. This recognizes the contributions of farmers, and identifies ways of protecting their rights, and of ensuring that benefits are shared with the countries in which they originated. The CBD established a legal framework for regulating access to genetic resources, and for benefit-sharing, and such legal and institutional frameworks will doubtless continue to evolve.

Whoever owns the genes, whether copyright or common heritage, they are a precious resource and need conservation *now*.

Genetic modification – solution or problem?

The issue of genetic modification (GM) has been much in the news in recent years, and there is a great deal of confusion about what it involves, and about its likely consequences.

Firstly it is fair to point out that genetic modification is a natural process involved in the evolution of life, indeed being central to it. Sexual reproduction produces variation through natural genetic modification. Down the ages people have changed organisms through selective breeding of species and varieties, altering domesticated animals and crops to suit their needs by improving yields and other desirable qualities.

More recently, refinements and advances in cell biology, genetics and biotechnology have led to new techniques for transferring genetic traits, and it is this technology that is generally referred to as GM. It involves transferring genes from one organism to another (which may be unrelated), thus adding particular traits to the recipient, and achieving a result unobtainable in nature or by traditional artificial selection. For example, genes from the bacterium *Bacillus thuringiensis* have been used to produce GM forms of a number of crops, including maize, oilseed rape, potato, rice, soybean and tomato. The modified crops produce a toxin that protects them from attack by insect pests. However, there is evidence that some of the insects develop resistance, and that some beneficial insects are also killed.

GM crops were first grown in the US in the late 1980s and early 1990s, and three-quarters of the world's GM crops are now grown in the US and Canada, the main crops involved being soybean and maize (corn) in the US, and oilseed rape in Canada.

Supporters of GM claim that this technique heralds a breakthrough in producing higher-yielding crops needing fewer herbicides, and that it will help solve the world's food shortages. The reality though is that damage to the environment may be a consequence of GM crop use. The herbicides designed to be used with the GM crops are very strong, and also the weeds they kill at first have in some cases been shown to evolve resistant strains ("super-weeds"), requiring the use of more and more herbicides. There is also some evidence that GM crop yields are no higher than those of more conventional varieties, and indeed are sometimes worse.

Another threat is to organic crops, as it seems GM and organic crops cannot co-exist, with the danger of genetic mixing between the two, thus destroying the organic status of nearby non-GM crops.

Trials of GM crops in the UK have concluded that they can be harmful to wildlife. These trials, which lasted nearly five years, ending in March 2005, involved planting "biotech" crops of oilseed rape, maize and beet in controlled plots. Weeds were then monitored, along with associated wildlife such as bees, butterflies, and other valuable insects.

One of the main reasons for this effect is that GM crops are designed to tolerate ultra-powerful weedkillers, which not only add potentially damaging chemicals to the soil, but which, by killing the other plants, including crop weeds, reduce the biodiversity.

The final test of GM crops found they caused harm to wild flowers, bees, and butterflies, and also probably to songbirds. The Royal Society for the Protection of Birds (RSPB) has published evidence of catastrophic declines in farmland birds over the past 50 years, and has suggested that growing GM crops would be very likely to make this even worse. The weedkillers, which are part of the GM crop package, kill plants on which birds such as tree sparrows, skylarks and bullfinches depend.

The American biotechnology giant Monsanto, the world leader in GM crop production, has had to withdraw from the European market in the wake of general protests, not just from environmentalists, but also from the growing lobby supporting organic crop production and healthy food. The future then for GM in Britain and in many other parts of Europe looks uncertain, and the balance seems to be shifting towards a more environmentally friendly future, which can only benefit the wildlife.

In the UK, the Soil Association is also concerned about the effects of GM crops on the health of people and on the environment, and, reflecting public concern, the main supermarkets have come out against using GM ingredients in their own-brand foods.

Despite these problems and concerns, GM crops are now grown in at least 18 countries, and the global commercial value of such crops was US$44 billion in 2003–04, with 98 percent being produced in five countries – US, Argentina, China, Canada and Brazil. The main crops concerned are soybean, cotton, maize and oilseed rape.

Recently, a new variety of GM rice ("golden rice") has been engineered. This produces about 20 times more betacarotene than previous varieties. Betacarotene is converted by the human body into vitamin A, a deficiency in which is a cause of childhood blindness. As many as 500,000 children go blind every year as a direct result of such a deficiency and it is hoped that this GM crop could help.

As mentioned at the start of this chapter, by the year 2025 it is estimated that food production will need to double to keep pace with demand. There is therefore a pressing humanitarian need to increase food supplies, especially in developing countries. Promoters of GM claim that this technology is the answer to the problem, but, as we have seen, it comes at a cost.

Another form of genetic modification involves adjusting the plant's own genetic "library" of information, rather than inserting "alien" genes. It is now becoming possible to awaken "sleeping" genes which can then express themselves in the plant. Examples are new forms of crops which can ripen in colder climates, or withstand salty conditions. This would enable them to be grown in a wider range of climates, and to be irrigated using salty water – even sea water. This can be seen as a more natural form of GM, and one that is potentially less hazardous.

Consumers wishing to avoid eating GM products face a tricky task, as GM soya meal finds its way into a huge range of foods, including margarine, bread, cereals, sausages and baby foods. Furthermore, soya is now the main nutritious feed for boosting

the production of cattle and poultry, and a good proportion of such feed is from GM soya. The expansion of the Brazilian soya market is also yet another threat to the rainforests of the Amazon, with large tracts of forest being destroyed, to be replaced by soya fields to service an expanding and lucrative export market.

The environmental view is that the genetic resources of local varieties of crops are far more valuable than the chemical-hungry monocultures represented by GM crops, and also far less damaging. Local landraces (traditional cultivated varieties) can improve the livelihoods of farmers by reducing the vulnerability of crops to disease and environmental changes, without the need for high-input pesticides and herbicides. In this context it is a tragedy that every year many traditional crop varieties disappear. In China alone, for example, almost 2,000 varieties of rice have been lost in the last 30 years.

10 October 2003: Greenpeace activists set up a mock corn-on-the-cob field in front of Berlin's parliament building, the Reichstag, to protest against genetically modified corn, **above**. *About 50 Greenpeace activists demanded the destruction of genetically modified crops to avoid the risk of contamination to the food chain.*

In South Australia, a farmer uses a field testing kit, **right**, *to see if his crop has been contaminated by Monsanto's now defunct 2001–2003 "Roundup-ready" canola trials.*

CHAPTER ELEVEN

Saving the Plants that Save Us

Not long after the Pilgrims had landed in North America in 1620, an early traveller described riding across prairies where "the strawberries grew so thick that their horses' fetlocks seemed covered in blood". What an evocation of an all but virgin land! A land whose bounty the Indians used so lightly that it regenerated when they had passed. Today only about one percent of the prairies remains as the settlers found it.

However, human influence on pristine environments goes back much further than this example. When Europeans first encountered Easter Island in the Pacific Ocean they found it barren, and they were also at a loss to explain how the inhabitants could have created the huge stone statues. But we now know, partly from pollen studies, that the island had a subtropical forest containing tall palms, and it was these trees that helped the people create the statues. Deforestation continued, the Easter Island palm became extinct around 1400, and the whole ecology of the island was severely damaged, streams dried up, and supplies of natural food disappeared, until the original society was all but destroyed.

In recent decades, the effects of modern agriculture on the countryside of Britain and Europe are also well documented, with flowers and birds that were once common becoming much rarer, or even extinct.

The world's natural vegetation is in deep trouble, and with it the soil that sustains humanity. The major cause of this ruination is "the paving over, digging up, ploughing under, overgrazing, chopping down, poisoning, flooding, burying, blasting and trampling of natural ecosystems". And the

Fire can spread very quickly, especially in dry habitats. Here, in Brazil, a brush fire has taken hold in the "cerrado" – a mosaic of scrub, dry grassland and open forest.

agents are people: ever-increasing population, with the consequent spread of habitations and the other human appurtenances like roads (cities are expanding and fertile ground disappearing at a rate of 65,000 sq. km per year), the need for food and all the other demands we make – for firewood, for animal fodder, for raw materials – plus a good deal of sheer greed.

Much emphasis has been placed on the tropical rainforests. In their amazingly complex ecosystems, the plant and animal life is unique. They may contain as many as half the world's stock of species. Many of us still think of the rainforests as huge, inviolable and everlasting. Alas, they are perhaps the most fragile, easily ruined ecosystems on planet earth. Estimates as to their potential survival vary from 20 to at most 50 years, if this destruction continues unabated.

As far as natural forests are concerned, losses in temperate regions, especially in Europe and North America, have been even more dramatic, with only about one percent of the original cover left in these regions.

Estimates of forest loss vary, but there is no doubt that natural tropical forests are particularly at risk. Global losses have been put at 160,000 km per year, or a total loss of 1,580,000 km between 1990 and 2000, a shrinkage of nearly nine percent in just ten years (World Resources Institute, WRI). Deforestation in the Amazon increased by 40 percent between 2001 and 2002, and much of this was to make room for expanding soya production.

In some areas the situation is even more acute. Brazil's Atlantic rainforest, which extended about 2,300 km in the nineteenth

Many aquatic plants are very susceptible to chemicals and very low pH caused by acid rain. Water soldier (Stratiodes aloides) is one such plant; it inhabits ponds and ditches, sinking to the bottom in its winter resting period, probably by accumulating calcium carbonate in the rosetted leaves. A local plant, it is diminishing in many parts of its range in Europe and North-west Asia because of drainage and water pollution.

century, is reduced to a number of totally separate fragments, covering not more than five percent of its former area; it is now the subject of major international rescue efforts. And most of the lowland forests outside protected areas of peninsular Malaysia and the Philippines have been logged. Forty years ago, who would have predicted that the rainforests of the world, once symbolic of remote places, would be in danger? The speed of loss is incredible.

The motive for destruction may be timber extraction – the demand for tropical hardwoods shows no signs of diminishing, let alone that for wood chips for paper, board, and chipboard – or clearance for cattle-ranching grassland, or crops of various kinds, including trees. Even if forest is logged and left, regeneration of secondary growth is a different quality: the trees of the mature forest may not grow again, if only because their seeds have very short viability, and it has proved almost impossible to nurture and plant out most of these trees of ancient lineage. Although logging has dramatic effects, in fact clearance for agriculture is the main cause of rainforest loss.

Island floras are particularly vulnerable. In Madagascar 80 percent of original rainforest cover has already vanished. To blame are slash-and-burn agriculture, frequently resulting in uncontrollable fires (quite often deliberately allowed to go as far as possible), and unbridled exploitation of trees and forest products. On smaller islands, human activities and the depredations of various animals – many of them introduced, either accidentally or deliberately for food, like goats (as on so many of the Galapagos Islands) – cause ever-increasing destruction of the often tiny populations of rare endemic plants and animals. The Mascarene Islands of Mauritius, Réunion, and Rodrigues in the Indian Ocean have a flora rich in endemics: Rodrigues for example has a total flora of about 135 taxa, of which about 65 are endemic to the Mascarenes, and nearly 50 endemic to Rodrigues. Sadly several are endangered and some have already become extinct. Introduced animals such as deer, monkeys and rats have affected the native species, and invasive plants such as Chinese guava have spread into the native vegetation. Similar situations exist on other islands, including Galapagos, Juan Fernandez, Seychelles, St Helena, New Caledonia and Hawaii.

Of course the rainforests and the island floras are not the only great natural systems at risk. Many northern woodlands have been turned into long-term tree monocultures, while the savannahs of the African Sahel countries have ecosystems as fragile as the rainforest.

Natural meadows and prairies are also becoming rare: there is hardly any meadowland left in Britain – it has either been made rich for grazing and poor for plant life by fertilizers, or has been ploughed up for crops. Hedgerows, admittedly "artificial", have largely been swept away, and with them a wonderful flora and fauna familiar only a generation back. Many wild flowers, once a common feature of the countryside, have become rare. Drainage of marshes and wet grasslands for agricultural improvement irrevocably destroys a further range of temperate flora. And most of the prairies of North America have been replaced by endless, enormous, cereal fields.

Even where forests are left more or less alone, the plants of the woodland floor can readily be over-grazed or over-exploited. The contraceptive-producing yam is almost extinct in the once forested Himalayas, the fabled ginseng endangered in China.

In areas that have long been cultivated in traditional ways, modern methods, machines and herbicides wipe out wild "weeds" and, more importantly for our own future, antique landraces and ancestors associated with crops.

The world's coastlines and estuaries are vulnerable too. Algal beds, coral reefs and coastal estuaries provide vital hatcheries for fish, crustaceans and molluscs; but we take too little care of them. The dredging of shipping lanes completely alters the seabed; silt from eroded lands damages reefs, smothering the fragile coral; sewage leads to blooms of marine algae which consume all the oxygen in the water. New York City for instance produces about five million tonnes of sewage sludge each year which used to be dumped at sea, causing major damage to marine life. Nowadays modern plants convert the sludge to pellet form, sterilized by heat treatment, and even used partly for fertilizing the land. Many coasts adjacent to big conurbations suffer pollution from sewage and also from industrial effluents.

Even the valuable mangroves, which not only protect tropical coasts but also nurse rich fisheries, fare no better. It has been plainly demonstrated that natural fish and prawn production in the vicinity of mangrove belts diminishes greatly if they are removed. Yet throughout the tropics, mangroves, a vital economic resource for local people, are in danger. Vast areas are being destroyed, either intentionally or as a secondary result of other activities. Mangroves also serve as a natural physical barrier to destructive high tides and tidal waves. They suffer from refuse and solid wastes often dumped in swamps, from river silt, mine wastes, sewage and chemical effluents, and from oil discharged by ships. Some are even removed to make way for shrimp farming; in fact the construction of ponds for this industry is one of the main causes of mangrove destruction. Shrimp farming is highly profitable in the short term, but upsets the local ecosystem. The ponds that replace the mangroves are stocked with shrimp larvae, which are fed with protein pellets, and pesticides and antibiotics are added to keep the over-stocked shrimps healthy. The water is pumped and replaced, polluting the sea nearby. But traditional systems of aqua-culture exist that do not harm the environment, such as integrated fish and rice farming common in rural China and elsewhere in Asia.

All over the world the pressures on wild plants, and more especially on large areas of natural vegetation, mount literally week by week. Too often their removal is carried out with no planning for maintenance of the denuded soil, with the result that we are soon faced with terrain that either is infertile, over-steep, too wet or too dry, or simply has no topsoil to sustain any plants at all, let alone useful ones. The consequences of such mismanagement may sometimes be irreversible.

It has been estimated that some 20 percent of the world's floras may be danger-ously rare or already under threat of extinc-tion. But if appropriate action is taken now a large proportion of these could be saved.

IUCN, the World Conservation Union, cat-alogues endangered species of animals and plants in its Red Data lists and books. The number of plant species recorded as threat-ened in 2004 stood at over 8,300, a figure which has steadily increased, and which certainly underestimates the true figure.

It is encouraging that most developed countries have prepared lists of their threat-ened plants as a guide and stimulus to action; it is less encouraging that few have done much to implement these gloomy findings. There is a tremendous role here for volunteers and activists, taking action at local level to save their own wild plants and natural heritage.

The hard fact is that in most tropical countries, especially those with rainforest, it is usually not possible even to identify individual threatened species. The plants are too numerous, too scattered in distribu-tion and too little known. In some cases they will become extinct before we even know they were endangered. Here the emphasis has to be on identifying not the threatened species but the sites where the most species can be saved.

One very useful initiative – Important Plant Areas (IPAs) – is being spearheaded by Plantlife International, the charity dedicated to conserving plant life in its natural habi-tats, in Europe and across the world. This approach focuses on places that are prior-ities for saving from the plant diversity point of view, and is more practical in many parts of the world than species-by-species approaches. IPAs are natural or semi-natural sites exhibiting exceptional botanical rich-ness and/or supporting an outstanding assemblage of rare, threatened and/or endemic plant species and/or vegetation of high botanical value. Plantlife International is co-ordinating the IPA programme in Europe, and is lead partner with IUCN The World Conservation Union for co-ordinating Target 5 of the Global Strategy for Plant Conserva-tion which aims for "protection of 50 percent of the world's most important areas for plant diversity assured by 2010".

One scientist estimates that for every plant species made extinct, from 10 to 30 other non-plant organisms may go with it. They depend on the plants for food, or some part of their life cycle (as indeed many plants depend upon animals). Large-scale forest or grassland destruction wipes out literally thousands of animal species of every type, large and small. Conservationists have mostly publicized the dangers of the extinc-tion of large, picturesque animals – giant panda, tiger and polar bear, for example. It is time to realize that every living creature depends on a habitat which is primarily constructed by plants. Today we see these

ecosystems being diminished, degraded or utterly destroyed. The rates of destruction and extinction are unprecedented in evolutionary history. In recent centuries documented extinction rates are several times higher than the "background" rate, or the rate at which species have been going extinct for the past 65 million years (that is since the last great extinction event at the end of the Cretaceous period when the last dinosaurs perished). Current extinction rates are nearly 1,000 times higher than "background" and set to climb still higher. Calculations such as this imply that a third to two-thirds of species could be lost during the present century unless habitat losses, particularly to the species-rich tropical forests, can be halted or reversed.

There is a deep-seated feeling among people that nature is ever bountiful and land ever plentiful. The less favoured, desperate simply to keep alive, can be forgiven for looking little further than their noses. But those in the "developed world" seem also extraordinarily reluctant to face facts. A flight over Toronto, to choose just one modern city, will demonstrate the lavishness of highways and enormous flyover junctions, while a run along the Queen Elizabeth Highway from Toronto to Niagara shows how the absence of legal constraint has resulted in ribbon development and a considerable loss of some of the area's most fertile land, a story repeated the world over.

Much of the destruction borders on the irrational. It has indeed been called subsidized vandalism. There seems to be something irresistible about the chainsaw. Away from forests, power machinery is equally damaging to vegetation. The impact of machinery on the world's wildlife in the past century, and most of all in the last 20–30 years, has to be compared with the greatest cataclysms of the geological past.

The species of plants being extinguished by our activities today represent the genetic heritage of aeons; their demise represents biological massacre. The destruction of forests and other natural habitats stops evolution dead in its tracks. As the great biologist Professor Edward O. Wilson has said, the event "our descendants will most regret is not limited nuclear war, nor energy depletion, nor economic collapse, but the loss of genetic and species diversity." Because this loss will take millions of years to compensate for – longer, perhaps, than the span of humanity itself.

Polluting the air

It has been recognized for a considerable time that factory chimney effluents are capable of damaging vegetation severely, but these effects have usually been considered relatively local. Today it is abundantly clear that this is not so – the effects can range far afield. Soot particles traced to

wood fires in India have been found on the Arctic ice cap, hastening its melting.

The words "acid rain" are now in common usage; they describe the damaging, polluted rainfall that results from the emission of sulphur dioxide and nitrogen oxides from chimneys and vehicles. In the air, these form sulphuric and nitric acids; and winds can carry them hundreds of miles, to fall as rain. The acidity of this rain is commonly around pH 4.6 – the same as orange juice; but even levels of pH 2.0 – stronger than vinegar – have been recorded.

Scandinavian countries have known for over a decade that acid rain, blown from Britain across the North Sea, is making their lakes so acid that there are now no fish in many of them. All aquatic life is affected by the damage inflicted on freshwater plants, particularly in those freshwater lakes and rivers on granite. Such waters, naturally low in mineral nutrients (especially calcium), have no buffer resistance for acid rain; the increased acidity quickly makes them unsuitable for a wide range of organisms.

Acid rain has damaged large tracts of forest in Norway and Sweden and in the Black Forest, Germany, where the proportion of damaged trees rose from 8 percent to 34 percent in 1982–83. Los Angeles smog has killed or damaged over a million of the Ponderosa pines in the surrounding forests. Governments generally are unwilling to lose tax revenue by enforcing remedial measures on the perpetrators: industrial chimneys and the motor car.

Acid rain apparently attacks tree foliage directly, and then has a further effect on the soil: the sulphates in the rain bind with

Lecanora conizaeoides

essential plant nutrients in the soil, such as calcium and magnesium, which then link up with water and leach away – becoming unavailable to roots. Magnesium deficiency causes leaf yellowing, and eventually acute poisoning sets in from aluminium, manganese and other elements that inhibit the division of cells in the roots of trees, so exposing them to disease pathogens. Tree death occurs through a combination of starvation, disease and poisoning.

Another aspect of acid rain is that it increases nitrogen fertilization – not always welcome in balanced natural ecosystems. Thus it has encouraged rank groves of stinging nettles in many sites in England and elsewhere in Europe, and encouraged for example the spread of *Molinia* grass in lowland heaths.

Atmosphere is further polluted by forest burning. Some parts of Brazil have atmospheric conditions resembling those of an industrial city because the smoke of burning trees produces photochemical reactions exactly like those of polluted smog.

Desertification

Besides the problems of pollution, there are the other effects on atmosphere which were outlined in Chapter 2. Destruction of forest and other vegetation is at the basis of a gloomy scenario, due either to the warming up of the globe because of increases in greenhouse gases or to the cooling albedo effect of greater reflection of solar radiation, caused by losses in vegetation cover. Reduced vegetation cover in areas of convectional rainfall (for example in the

Lichens are unique in being composed of two organisms, algae and fungi, which form a close symbiotic association: the fungus greatly benefits by the alga's ability to produce chlorophyll and thus to photosynthesize. In return fungal tissue protects the alga from intense light, drought and heat. Together they are able to colonize harsh environments. But lichens' efficient absorption mechanisms make them highly susceptible to contamination – especially by sulphur dioxide. Their presence or absence is therefore a useful indicator of atmospheric quality.

The map (below) shows increasing lichen-richness away from industrialized areas. The unit of pollution is the mean winter level of sulphur dioxide in microgrammes per cubic metre of air.

■ Only crustose lichens, if any
Heavy pollution (150 units)

■ Crustose lichens and a few foliose
lichens (70–125 units pollution)

■ Lichen flora increasing
(60 units pollution)

□ Beard lichens occasionally found
Pollution low (40–50 units)

■ Rich lichen flora
(35 units pollution)

■ 25 species plus on large trees
Pure air (under 30 units pollution)

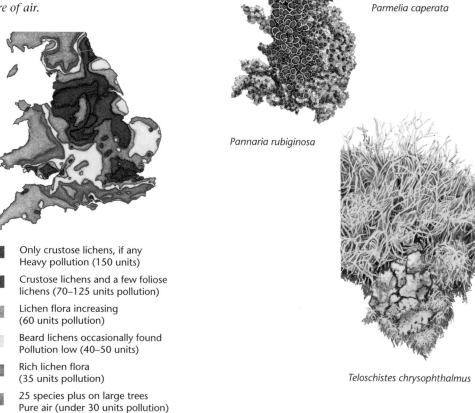

Parmelia caperata

Pannaria rubiginosa

Teloschistes chrysophthalmus

Madagascan workers walking along a dry sisal plantation road during the drought of 1985–86. The massive dust dispersion was caused by desertification.

tropical forested regions) results in lower evapotranspiration, lower levels of recycling of water through the atmosphere, and reduced rainfall downwind.

Thus the destruction of forest, and corresponding reduction in water vapour transpired, reduces the amount of rainfall in a given area in an often drastic manner. The effects just mentioned may also induce large-scale climatic change.

One of the places where this appears to be occurring is the northern part of Africa, especially the Sahel belt including the countries of Mauritania, Mali, Niger and Chad. The southern edge of the Sahara Desert has in recent years crept southwards; rainfall has decreased and in particular been much less regular, often arriving unseasonally, and even missing whole years. Desertification here has been accelerated by human activities – with famine, hardship, and the disruption of wars and refugees all playing a role.

Lands become unnaturally barren in several ways. If too many animals are grazed on sparse rangelands – not only cattle but also goats and thorn-relishing camels – they destroy the plant growth, except a few remnants of inedible scrub, and trample the soil into dust. Coupled with this shortage of forage is shortage of fuelwood, so people strip and cut down

trees in ever-widening circles from their settlements. Such activity leads to the eventual loss of topsoil, due to wind and the occasional torrential downpour which causes a sheet of water to move across flat areas carrying the precious soil with it.

Another certain way is by ploughing up and down steep slopes without terracing; this puts the topsoil at risk in heavy rain. Slopes of over 35 percent gradient should not be cultivated without special protective measures. And, further, barrenness can follow in the wake of ill-conceived irrigation schemes which deposit salts, creating a saline wasteland.

Desertification as a result of vegetation destruction without replanting is occurring in many places; some on the edge of existing natural deserts like the Atacama, Sahara, Sonoran, Great Indian, Kalahari, and the Taklamakan and Gobi Deserts of China and Mongolia. Desertification also affects patches of arid land far from the natural deserts. One quarter of the earth is threatened by desertification, according to UNEP (United Nations Environment Programme) estimates, and this in turn impinges on the livelihoods of at least a billion people in more than 100 different countries. An area roughly the size of China and India (1.2 billion hectares) has suffered soil deteriora-

tion in the last 50 years, mainly in arid or semi-arid regions in developing countries.

Present day desertification and, indeed, destruction of natural vegetation in general, worldwide, are inextricably tied up with the increase in global population, combined with poverty and over-exploitation and unsustainable use of natural resources.

There is more than enough food produced worldwide to feed everyone, but at present enough food aid is most certainly not reaching most of the billion or more people who are at risk of starvation. Almost half the population of Sub-Saharan Africa are affected by poverty or malnutrition. And as long as people are destitute they will, as in Africa, continue their reliance on local resources or, as in Amazonia, lay waste primary forest in order to plant subsistence crops, while those better off try to raise cash from, for example, beef cattle, oil palms, or bananas.

The spread of ranches and plantations, and also Green Revolution agriculture displaces poorer rural people, who then settle in agriculturally marginal areas, thus causing further habitat destruction.

A reduction in the rate of growth of the human population is desirable for adequate comfort and nourishment, as well as for the conservation of what is left of our green inheritance.

The world's population currently stands at about 6.5 billion, and is estimated to climb towards about 7 billion in the next ten years, and 9 billion by 2050. If the rates of land degradation also continue, how many more people will then be destitute? The good news is that the annual rate of

Franklinia alatamaha, **above**, *once found by the Alatamaha River in Georgia, now survives only in cultivation.*

The African violet (Saintpaulia ionantha), **below**, *symbolizes WWF's international plant conservation efforts because it is severely threatened in its East African jungle home, a habitat of steep rock surfaces in dense shade, endangered by forest clearance.*

increase has slowed in recent years, from a peak of over two percent 40 years ago, to about 1.1 percent in 2005, and that this rate is set to continue a downward trend. The sad news is that this decrease is only partly due to a decline in the birth rate, but is also a consequence of the ravages of disease, and notably of HIV/AIDS, especially in Africa, where life expectancy can be as low as 30 years.

Solutions

There are two main approaches to saving plant cover, both urgently needed. The first is to halt degradation and rehabilitate land already damaged, so those who live there can continue to use it. The second is to conserve as much as possible of the remaining natural (and semi-natural) vegetation – not just as a living museum, but as a resource which can provide for the future.

Some remedies to prevent degradation are glaringly obvious. Protecting the air from factory effluents can be achieved by installing filters; it can be done – at a cost. Protecting water is, equally, an expensive matter of curbing the dumping of untreated effluents and oil.

In poor countries suffering desertification, the solutions are again obvious in principle. Trees and other growth can be protected by reducing the number of grazing animals, or by fencing areas for new plantation or regeneration. To stop erosion on slopes, if cultivation terraces are too difficult to create, the simplest remedy is to plough along contours, and dam gullies to trap water in heavy rain. Sheet erosion can be prevented

by ridges as little as 20 cm high across gentle slopes, while a semi-circular ridge on the downhill side of newly planted trees traps water, and greatly increases their chances of survival.

Reafforestation, by the planting of seedling trees after clear-felling, is an accepted part of the logging cycle in industrialized countries, but is less often practised in tropical forests. Holders of felling concessions may complain that their leases of only 15–20 years are too short to provide incentive, but in fact most foreign logging companies like to extract timber as fast as possible. The loggers are not solely to blame; timber rights are let to the highest bidder, without control or direction for reinstatement by governments forced to take short-term actions to satisfy immediate human needs – food, shelter, water.

In some places, selective logging has been a deliberate policy in the expectation that this would not harm forests materially, and would allow them to regenerate. In practice, it is usually impossible to avoid severe damage; even if only a few large trees are taken here and there, the trails needed for getting access to them and removing them open up remote parts of forests to poaching and illegal exploitation of trees and minerals, which is equally destructive. Some tropical trees are known to suspend their growth for up to 20 years if disturbed: no one yet knows why, or how. Any removal of trees also alters the species composition of the forest, which is almost always detrimental in the long term. In principle, such disturbance usually devalues or ruins tropical moist forest.

181

Tree plantations are an obvious improvement for logged or otherwise degraded land, and many countries have schemes in operation; the Chinese government, for instance, funds the growing of young trees in nurseries, and their distribution; and local people then plant and tend them, in some areas as an attempt to reverse desertification.

It is expected that tens of millions of hectares of plantation will be established in the tropics in the coming years – though deforestation at the current rate will exceed this replacement. Many kinds of trees are used, including arid-tolerant acacia species, notably *A. albida*. Eucalypts are the classic fast growers, and lamtoro (*Leucaena leucocephala*) is another contender. It grows very rapidly to a height of 20 metres and each hectare can produce 30–40 cubic metres of biomass a year. Out of the 42 million hectares in Indonesia needing re-greening, lamtoro is reported to have rehabilitated 3.7 million hectares in four years. Its merits

include controlling erosion and resisting drought; providing forage, firewood of high fuel value, timber, windbreaks and shade; and improving the soil (being leguminous it has nitrogen-fixing bacteria on the roots). Despite all such merits, plantation trees are much less efficient than big forest trees in producing oxygen and, more importantly, transpire only from one-tenth to one-fifteenth as much water, so rainfall in replanted areas, though it will improve, will never be as much as over rainforests.

Parks and reserves

By far the biggest threat to plants is that the habitats in which they grow – rainforest, desert, savannah – are themselves in danger. So a top priority is to build a network of protected areas – national parks and nature reserves – covering representative samples of the best habitat types. These will protect populations of species and also serve for

The lower reaches of the Gordon and Franklin rivers were in the news in the early 1980s, as the Tasmanian government had plans to flood this area of considerable biological and archaeological importance. Conservationists were in uproar, especially as the area had been declared a World Heritage site, nominated by the Australian federal government for protection under the Unesco World Heritage Convention. TV botanist David Bellamy was incarcerated for his protest – to good effect as the dam was eventually banned. This area is one of the last remaining temperate rainforests, dominated by southern beech (Nothofagus). Two-thirds of the island's endemic plants live there and 18 species would have been threatened if the scheme had gone ahead. Among these is the Huon pine (Dacrydium franklinii) – 35 percent of its habitat would have been flooded.

scientific and historical study – future generations will only have these to appraise. They may be the only areas of true wilderness left on earth. Ideally too, many such reserves will include buffer zones in which the local people can continue to exploit their traditional natural resources sustainably.

Above all they will protect wild genetic resources, which as we have seen throughout this book, are so important to human welfare. The present view of conservation is a change in emphasis from conservation for its own sake to conservation for the benefit of people. Conservation bodies such as WWF, Plantlife International, and People and Plants International are now making special efforts to safeguard plants used by people, whether or not these enter the monetary economy of governments. In so doing, it is hoped such conservation will also support those peoples that still live in harmony with their environment, with their invaluable knowledge, and maintain habitats for the innumerable animals that would otherwise vanish with the plants.

Along with this is a recognition that conservation organizations, with limited funds, can achieve most not by concentrating on single species, but by focusing on a type of habitat, like the rainforest. Because vegetation is being lost so fast and because of the sheer number of species threatened, it is neither possible nor sensible to plan a reserve for each species.

Clearly a priority is to find those areas with the most diversity, and protect them. Setting up one big rainforest reserve could alone save hundreds, if not thousands, of species. In some cases, the areas are those believed to have been the refugia of tropical vegetation during the late Pleistocene epoch, when much drier conditions than today's affected the tropics, and the forest retreated into small pockets which became centres of species evolution. In the Amazon basin, for example, Brazil has based its protected area plan upon studies of these Pleistocene refugia, which are the chosen areas for the big parks Brazil is creating. Yet the impressive total of ten million hectares declared in Brazil is only 1.2 percent of that vast country. The Global Strategy for Plant Conservation, agreed in 2002, aims, among other targets, to assure protection of 50 percent of the most important areas for plant diversity (see page 186 for targets).

In some cases a single species acts as symbol and rallying point for a whole programme of habitat conservation. Project Tiger, which WWF-India supported with a major campaign, led not only to an increase in tigers from about 1,800 to over 3,000 but, even more important, to a revitalizing of India's protected areas network, with benefits to all the numerous plants and animals with which the tiger shared its habitat. Of the estimated 3,000–3,500 tigers in India today, only about a third are in protected reserves, the remainder being dispersed and more vulnerable, and illegal poaching also occurs occasionally even in the reserves. Similarly, the reserves set up primarily to protect the giant panda in China, can only succeed by protecting large areas of pristine habitat, along with many other endangered species of animals and plants.

More recently, efforts have been made to promote Joint Forest Management, notably

The ginkgos – primitive, deciduous conifer relations – came into being about a quarter-million years ago and fossils confirm their worldwide distribution. The present day maidenhair tree (Ginkgo biloba), little changed from its ancestors, was last seen in the wild in the Tian-Mu range in China, but was reintroduced to the world in the eighteenth century from Chinese and Japanese temple gardens. Possibly wild examples are still known from certain mountain forests in central China. It was grown for its unique foliage, stately growth habit and longevity, and also for its edible seeds. Now it is widely planted in parks and gardens and as a street tree in many North American towns. Mostly these are a narrow upright form known as "sentry" which, being male, have no fruits; but in some places female trees have been used and their malodorous seeds are a menace in autumn. In towns, ginkgos put up with the aridity of pavement planting and severe air pollution with such tolerance to inimical conditions it is a mystery why they almost became extinct.

in India, where People and Plants has been active for example at Ayubia National Park. There is increasing realization of the importance of forests for local livelihoods and that local communities need to be involved in their management in a joint approach, sharing in decision-making, to the benefit both of the local people and of conservation.

Yet it has to be realized that the creation of protected areas is principally a means of buying time. There cannot be effective conservation in the middle of over-populated, poverty-stricken and largely hungry agricultural communities; the pressures and temptations are too great when protected areas become lush – but forbidden – pockets of vegetation, in a sea of degraded and transformed land. This underscores the importance of the people-friendly approach to conservation, with locals encouraged to use the natural resources in (often traditional) sustainable ways.

Concerned at this development, enlightened park chiefs are changing their tactics. They understand that in developing countries, meeting human needs – food, health and shelter – has to be the primary goal. Rather than set the area aside, they want to protect it so that its benefits continue to radiate out into the surrounding countryside – migrating meat for the pot, fresh water in the streams, markets for local handicrafts. A key concept is that of the buffer zone, a broad margin between the park and the neighbourhood; here people can gather firewood and wild fruits, graze limited numbers of cattle, and perhaps harvest medicinal herbs – many local people feel that saving their valued medicinal plants is

perhaps the best justification of all for having national parks.

Ever since the United States declared Yellowstone National Park in 1872, parks have been created all over the world. In the decade between 1972 and 1982, major protected areas, excluding the smallest, rose from 212 million hectares to around 386 million hectares – an impressive five percent increase, and today there are more than 100,000 protected areas covering over 18 million sq. km. Yet even this covers only a small proportion of the earth's surface, at a time when vegetation is rapidly being destroyed or degraded. But encouragingly, there has been a big increase over the last decade, and the total of the planet's land surface under protected status now stands at nearly 12 percent, and protected areas now cover more land than do permanent arable crops.

Desperately needed are the skills and resources to create and manage protected areas, and to implement the concepts of the park in harmony with its neighbourhood and making a contribution to national development. Worldwide, park managers form a community of conservationists who have perhaps the most important and urgent job of all. Money is important, but trained and skilled manpower is even more vital. One of the characteristics of the conservation movement is how one persistent person on the ground can safeguard an area far better than any intervention from outside or any passing interest. Intelligence, persistence and above all enthusiasm – with these one can work miracles.

This has been the approach of the People and Plants initiative of WWF, UNESCO and

Royal Botanic Gardens Kew, which encouraged such approaches and which has bequeathed a legacy of case studies and on-the-ground schemes based on this co-operative ethic. This is also the philosophy of People and Plants International (PPI), which was founded in 2004 and which builds on the work of People and Plants. PPI is a global network bringing together local and international experts to combine traditional knowledge and science in integrated development, conservation and education projects. Its aim is to create sustainable local solutions to improve relations between people and the natural environment, recognizing that conservation and the survival of local people are intimately connected.

The place of botanic gardens

There are more than 2,000 botanic gardens worldwide, and over half of these make an important contribution to plant conservation. Together they maintain over 80,000 species in cultivation, and some also keep plants in seed banks and tissue culture. Of the 34,000 or so threatened plant species, over 10,000 are in botanic garden cultivation. Even a few species believed extinct in the wild – like the toromiro (*Sophora toromiro*), Easter Island's only tree – have been maintained in cultivation in several botanic garden collections.

As far as the public goes, these gardens, many near large towns, are the only places where large numbers of people can see exotic and unfamiliar plants. They are ideal places to tell the public about conservation and what people can do to help, especially where gardens have museums and exhibition centres. In many ways they are the shop windows of plant conservation.

These institutions can undertake rescue operations to places with many threatened endemics, where habitat conservation is difficult or impossible, like Socotra and Somalia, and bring examples of the plants into cultivation. Botanic gardens can save seeds in gene banks, nurture plants that cannot be cold-stored, and increase stocks of endangered species. Ideally every rare and endangered plant should be cultivated in at least one garden, and their genetic material should be in at least one seed bank; that is a minimum. With modern methods of propagation, like meristem and tissue culture, endangered plants can be increased for dissemination to other gardens.

Perhaps even more important with especially desirable ornamental plants, commercial markets can be flooded with specimens at low prices that will undercut the trade in wild-collected, often poached plants. Gardens can equally introduce, propagate and disseminate plants of economic importance. Botanic gardens are now co-ordinated as an ex-situ network for threatened plants and to encourage them to take up this new role. The main player here is Botanic Gardens Conservation International (BGCI), based near Kew Gardens in London, UK.

Experts can correlate existing strategies for conservation and regeneration, and experiment in new ones. Well co-ordinated data-collecting methods will vastly increase information capacity, as botanic gardens communicate with each other and build up knowledge worldwide. In this connection, one should not underestimate the existing storehouse of plant species in public and private gardens.

A final word

Ahead of us is one of the most difficult tasks which we have ever faced – comparable with that of avoiding a nuclear holocaust. Nations need to agree, and act together to curb the unbridled destruction of the remaining tracts of natural vegetation – stop it altogether if possible. It will be a hard struggle indeed for people of such different needs, wishes, outlooks and backgrounds all jostling with their own points of view – let alone for politicians, who seem never to consider any moment but the present. But if the destruction continues unabated, the earth will become largely covered first with crop plants and then, probably, barren lands, with catastrophic effects on the life-support system of planet earth and the future well-being of its inhabitants.

The pragmatic new attitude of international conservation efforts, such as those for example of the WWF and the World Conservation Union, towards plants as resources must surely make governments think again about conservation priorities. We are making a start towards saving plant life – Saving the Plants that Save Us.

Few will deny that there is a moral issue here: a simple enough axiom that we should not, for our own immediate comfort and gain, destroy plant life so wantonly and with so little thought for the future. For ourselves, and how much more for our descendants, there is a pressing need to leave something at least of the world we inherited, its plants and the animals they support, for the future to comprehend, enjoy and benefit from.

It is surely possible for us now to strike a balance. Technologically we have the means. It is incredible, really, to consider that so many billions of dollars are being spent on space research, that we know more about the moon than some parts of the Amazonian forests. Is it not time to divert some of these scientific and financial resources to our biosphere? The planets and stars will still be there in 20 years; tropical forests may not.

We now have the technological capacity to plant those land areas already modified for agriculture with sufficient crops, of every conceivable kind, to feed, house and clothe everyone well – making it unnecessary to destroy what remains of natural vegetation. Besides thus providing for our physical needs, plants can also provide solace for the mind. They are endlessly fascinating, exciting, wonderful – even fabulous is not too strong a word, and they are diversely beautiful too. And this fascination is greatest when one can see plants interacting with each other in the intricate web of life which botanists call an ecosystem, adapting themselves to every habitable niche on the earth's surface.

It is surely time we sealed a non-aggression pact with these remarkable planet-sharers and lived with them in a harmonious relationship.

A Future for Plants

Individual plants feed our world, cure our ills, provide materials for industry, and enrich our spiritual lives. They also protect fragile soils from erosion, and help to regulate the atmosphere, maintain fresh water supplies and prevent desertification.

Key facts

▶ Forests hold the largest number of plant species, yet nearly half of all forests have disappeared, and only about eight percent of the remaining forests are protected.

▶ Nine percent of the world's estimated 100,000 tree species are under threat, as are 1,800 types of orchid.

▶ More than 35,000 species of plant are used for medicine, including at least 7,500 in India.

▶ Plants have created a quarter of all prescribed medicines.

▶ Local crop varieties provide security for rural people, yet in Indonesia, to take just one example, 1,500 local varieties of rice have become extinct in the last 15 years.

▶ Over 3,500 tonnes of African cherry (*Prunus africana*) bark are removed each year from African forests and exported for a herbal trade worth an annual £140 million.

▶ Sales of herbal medicines have soared in recent years, leading to over-exploitation of many medicinal plant species.

▶ Rare tropical hardwood trees are felled to supply the woodcarving trade, degrading forests and endangering the species concerned.

Key aims of conservationists

▶ Conserve biodiversity to provide habitats for wildlife, and for ecological services, such as water provision and climate moderation.

▶ Halt and reverse deforestation.

▶ Find a balance between conservation and the use of wild plants to ensure that the livelihoods of harvesters are sustained, and resources available for future generations.

▶ Ensure that, wherever plants are exploited for commercial gain, the benefits are shared fairly, for conservation and local livelihoods.

A global strategy

A milestone in international plant conservation was the agreement, in 2002, to the Global Strategy for Plant Conservation under the Convention of Biological Diversity. This sets targets at global level, to be met by 2010. Community-based conservation and capacity-building are recognized as vital for its achievement, especially in the case of developing countries.

Key targets

▶ At least 10 percent of each of the world's ecological regions effectively conserved.

▶ Protection of 50 percent of the most important areas for plant diversity assured.

▶ At least 30 percent of production lands managed consistent with the conservation of plant diversity.

▶ 60 percent of the world's threatened species conserved in situ.

▶ 60 percent of threatened plant species in accessible ex-situ collections, preferably in their country of origin, and 10 percent of them included in recovery and restoration programmes.

▶ 70 percent of the genetic diversity of crops and other major socio-economically valuable plant species conserved, and associated local and indigenous knowledge maintained.

▶ No species of wild flora endangered by international trade.

▶ 30 percent of plant-based products derived from sources that are sustainably managed.

▶ The decline of plant resources, and associated decline of local and indigenous knowledge, innovations and practices, that support sustainable livelihoods, local food security and health care, halted.

▶ The importance of plant diversity and the need for its conservation incorporated into communication, education and public awareness programmes.

▶ The number of trained people working with appropriate facilities in plant conservation increased, according to national needs, to achieve the targets of this strategy.

▶ Networks for plant conservation activities established or strengthened at national, regional, and international levels.

Biodiversity mapped

Important sites for plant diversity have been recognized, notably the biodiversity hotspots identified by Conservation International (CI) and the Global 200 of WWF.

Biodiversity hotspots: Based in Washington DC, USA, CI aims to conserve the earth's living natural heritage and global biodiversity, and to show that humans can live harmoniously with nature.

A hotspot must contain 1,500 species of vascular plants (>0.5 percent of the world's total) as endemics, and have lost at least 70 percent of its original habitat.

Over half the world's plant species (at least 150,000) and over 40 percent of terrestrial vertebrates (nearly 12,000) are endemic to the 34 hotspots currently identified by CI.

Global 200: A ranking of the earth's most biologically outstanding terrestrial, freshwater and marine habitats. It aims to ensure that conservation efforts around the world contribute to a global biodiversity strategy and reflects three major innovations:

▶ **It is comprehensive in its scope.** It encompasses all major habitat types, including freshwater and marine systems as well as land-based habitats. It ranges from arctic tundra to tropical reefs, from mangroves to deserts, to include species from every major habitat type on earth.

▶ **It is representative in its final selection.** The most outstanding examples of each major habitat type are included from every continent and ocean basin. Thus it includes, for example, the most important tropical and temperate forests from each continent, and the most important coral reefs from each ocean.

▶ **It uses eco-regions as the unit of scale for comparison and analysis.** Eco-regions are large areas of relatively uniform climate that harbour a characteristic set of species and ecological communities. By focusing on large, biologically distinct areas of land and water, Global 200 sets the stage for conserving biodiversity.

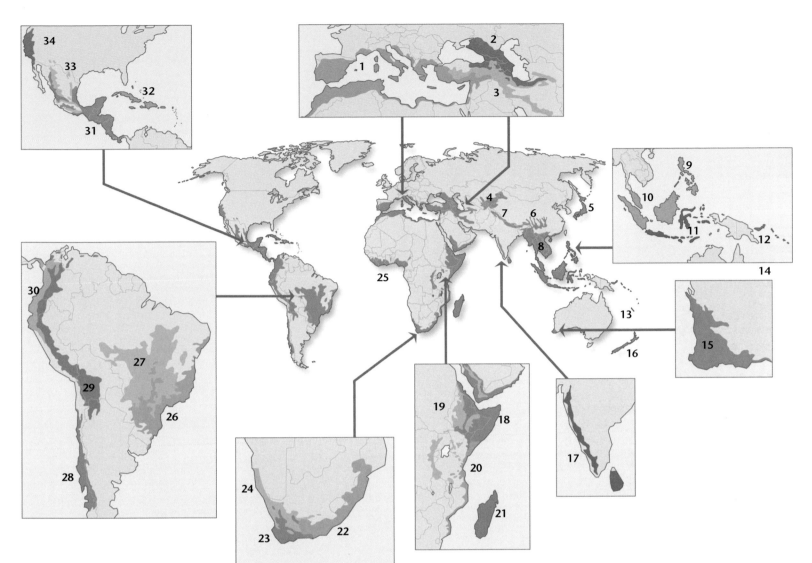

Biodiversity hotspots

*Conservation International has recognized biodiversity hotspots around the world. The 34 sites currently identified, **above**, are particularly rich in biodiversity, containing high concentrations of threatened plants and animals, and therefore especially worthy of conservation.*

1 Mediterranean basin
2 Caucasus
3 Irano-Anatolian
4 Mountains of Central Asia
5 Japan
6 Mountains of South-west China
7 Himalaya
8 Indo-Burma
9 Philippines
10 Sundaland
11 Wallacea
12 East Melanesian Islands
13 New Caledonia
14 Polynesia and Micronesia
15 South-west Australia
16 New Zealand
17 Western Ghats and Sri Lanka

18 Horn of Africa
19 Eastern Afromontane
20 Coastal forests of Eastern Africa
21 Madagascar and the Indian Ocean Islands
22 Maputaland-Pondoland-Albany
23 Cape Floristic Region
24 Succulent Karoo
25 Guinean forests of West Africa
26 Atlantic forest
27 Cerrado
28 Chilean winter rainfall-Valdivian forest
29 Tropical Andes
30 Tumbes-Choco-Magdalena
31 Mesoamerica
32 Caribbean islands
33 Madrean pine-oak woodlands
34 California Floristic Province

Further reading

This is just a small selection of recent books which may be of interest to readers wishing to find out more about some of the topics covered. In addition, many other publications, articles and websites were used in researching the new material for this edition.

Baillie, E.M. et al. (eds) (2004) *IUCN Red List of Threatened Species*. IUCN, Gland, Switzerland and Cambridge, UK.

Balick, J.B. and Cox, P.A. (1997) *Plants, People, and Culture: The Science of Ethnobotany*. Scientific American Library, New York, USA.

Lewington, A. (2003) *Plants for People*. Eden Project Books. Transworld Publishers, London, UK.

Lomborg, B. (2004) *Global Crises, Global Solutions*. Cambridge University Press, Cambridge, UK (technical)

Luhr, J.F. (ed.) (2003) *Earth*. Dorling Kindersley Limited, London, UK.

Mabberley, D.J. (1997) *The Plant-book: A Portable Dictionary of the Vascular Plants*. 2nd edn. Cambridge University Press, Cambridge, UK.

Mackay, R. (2002) *The Atlas of Endangered Species*. Earthscan, London, UK.

Marinelli, J. (ed.) (2004) *Plant*. Dorling Kindersley, London, UK.

Meyers, N. (ed.) (2005) *Gaia Atlas of Planet Management*. Gaia Books, London, UK.

Mittermeier, R.A. et al. (2005) *Hotspots Revisited*. Conservation International, Washington, USA.

Vaughan, J.G. and Geissler, C. (1997) *The New Oxford Book of Food Plants*. Oxford University Press, Oxford, UK.

The People and Plants Conservation Series. Earthscan, London, UK:
 Campbell, B.M. and Luckert, M.K. (eds) (2001) *Uncovering the Hidden Harvest*
 Cronk, Q.C.B. and Fuller, J.L. (2001) *Plant Invaders*
 Cunningham, A.B. (2000) *Applied Ethnobotany*
 Cunningham, A.B., Campbell, B.M. and Belcher, B. (2005) *Carving out a Future*
 Laird, S.A. (ed.) (2000) *Biodiversity and Traditional Knowledge*
 Martin, G.J. (1995, 2004) *Ethnobotany*
 Shanley, P. et al. (eds) (2001) *Tapping the Green Market*
 Tuxill, J. and Nabhan, G.P. (2001) *People, Plants and Protected Areas*

Useful organizations

Consult websites for up-to-date information

ARKive: Images of Life on Earth www.arkive.org
Center for International Forestry Research (CIFOR) www.cifor.cgiar.org
CITES (Convention on International Trade in Endangered Species of Wild Flora and Fauna) www.cites.org
Conservation International www.conservation.org
Convention on Biological Diversity (CBD) www.biodiv.org
Eden Project www.edenproject.com
Fairtrade Labelling Organizations International www.fairtrade.net
Fauna and Flora International www.fauna-flora.org
Forest Stewardship Council www.fsc.org
International Plant Genetic Resources Institute (IPGRI) www.ipgri.cgiar.org
IUCN (The World Conservation Union) www.iucn.org
Oxfam International www.oxfam.org
People and Plants International www.peopleandplants.org
Plantlife International www.plantlife.org.uk
Ramsar Convention on Wetlands www.ramsar.org
Royal Botanic Gardens, Kew www.rbgkew.org.uk
TRAFFIC (Wildlife Trade Monitoring Network) www.traffic.org
UNESCO MAB (Man and the Biosphere) www.unesco.org/mab
World Conservation Monitoring Centre (UNEP-WCMC) www.unep.wcmc.org
World Resources Institute www.wri.org
WWF International www.panda.org
WWF-US www.worldwildlife.org
WWF-UK www.wwf.org.uk

Index

Martin Walters' acknowledgment to 2nd edn
I should especially like to acknowledge the advice of Dr Alan Hamilton, formerly head of the International Plants Conservation Unit of WWF-UK, and now at Plantlife International. As one of the core team members, he was deeply involved in the influential People and Plants programme of WWF, UNESCO, and Kew, for which I was editor. I am pleased to have worked for many years with him and others in international ethnobotany, one of the main themes of this book.

Author's acknowledgments This book has been very much a collective effort and I gratefully acknowledge the immense help provided by the editorial team especially Joss Pearson. Richard Gorer and Juliet Bailey helped greatly with research on several chapters. The basis for the account of North American Indian ethnobotany came from the book *Plants in British Columbia Indian Technology* by Nancy J. Turner, who has kindly permitted the direct quotation made on p. 151. Quotation has also been made briefly from *The Law of the Seed* by P.R. Mooney (in Development Dialogue, 1983, 1–2) which provided invaluable background material for Chapter 10. I must also thank my wife Alyson who typed most of my initial handwritten draft.

Publisher's acknowledgments Gaia gratefully acknowledge the help of the World Wildlife Fund, the International Union for Conservation of Nature, and the Royal Botanic Gardens, Kew. Their advice, assistance, and provision of access to research data and the library and art collection at Kew were invaluable. In particular, we would like to extend special thanks to the following people: Paul Wachtel and Peter Palmer for all their support; Hugh Synge, Jeremy Harrison, Steve Davis, Stephen Droop and all the staff of the IUCN Conservation Monitoring Centre at Kew; Gren Lucas, Sylvia FitzGerald, John Flanagan, Marilyn J. Ward, Hilary Morris, Tudor Harwood, Milan Svanderik and all the staff of the library at Kew. Dr Paterson and Professor David Hall of King's College, Amory Hubbard and Linda McMahon of WWFUS, Libby, Juliet, Bridget, Tony, Imogen, Elly and Anita for all their work; Aardvark

Editorial for project managment and Bryony Allen and Linda Norris for the index.

Photographs Ardea London: 182, 178 (Jean Paul Ferrero), 116 top left (A. Green Smith), 98 bottom (Francois Gohier), 14 bottom right (John Mason). Aspect: 124 (J. Matthews). BBC Hulton Picture Library: 150 left. Biofotos: 13 left, 16 top, 51 bottom right, 135 top (Heather Angel). Botanica/Photolibrary.com: 76 top. British Museum: 58 bottom left, 134 left. Stefan Buczacki: 96, 98 top, 100 top (Vietmeyer). M. Clements: 163. Bruce Coleman Ltd: 22, 127, 116 top right, 147 top (Eric Crichton), 65, 70 bottom right (Cameron Davidson), 88 (Nicholas Devore), 14 top (M.P.L. Fogden), 61 top (Michael Freeman), 34 left (C.B. Frith), 146 bottom (Jennifer Fry), 62 (David Hughes), 61 bottom (O.A.J. Mobbs), 156 (Norman Myers). Corbis: 73, 76 (Bettmann), 80 (Lindsay Hebberd), 92 (Owaki-Kuila), 157 (Tom Bean), 159 (Oswaldo Rivas/Reuter), 173 top right (Tobias Schwartz/Reuters). Eric Crichton Photos: 141 top. Mary Evans Picture Library: 87 top. Getty Images: 10 (Robert Harding World Images), 160. C. Grey-Wilson: 14 bottom left, 16 bottom right, 33, 154 right. Susan Griggs Agency: 20, 56, 70 top and bottom left, 153 (Victor Englebert), 130 (Adam Woolfitt). David Hali: 78, 94. Robert Harding Picture Library Ltd: 37 top, 55 (C. Grey-Wilson). John Hillelson Agency: 36 (George Gerster). Michael Holford: 63 top right, 110 bottom left, 151 top right. Anthony and Alyson Huxley: 44 bottom, 47 left, 58 bottom right, 83 left, 87, 134 right. Mansell Collection Ltd: 74, 110 top left, middle, bottom, top right, bottom right, 111 bottom. Minden Pictures/FLPA: 174 (Claus Meyer), 180 (Frans Lanting). Tony Morrison: 27, 30 bottom, 31, 32, 100 bottom left, 105 left, 117 bottom, 152. Natural Science Photos: 51 bottom left (H.E. Axel), 85 top left (P.J.K. Burton), 115 bottom right (M. Chinery). Nature Photographers Ltd: 37 bottom (Christine Osborne). OSF Picture Library: 51 top (Liz and Tony Bomford), 176 (G.I. Bernard), 13 right (Sean Morris). Joss Pearson: 16 bottom left. Ann Ronan Picture Library: 44 top left, 48 bottom left, 49, 50 top. A.D. Schilling: 154 left and middle. Seaphot:

30 top (Ian Redmond). R. Sheridan: 63 middle right, 111 top. Harry Smith: 135 bottom, 138 bottom, 139 bottom, 140 bottom. Still Pictures: 29 (Klaus Andrews), 38 (Shehzad Noorrani), 52 (IFA), 173 bottom (Martin Wyness). Tony Stone Worldwide: 65. Survival Anglia/OSF: 44 top right, 109 top (Cindy Buxton and Annie Price), 109 bottom (Jeff Foott). Vision International: 132 top (D. Warren). Martin Walters: 103, 108, 128 (Tony Cunningham), 129 (Susanne Schmitt), 148. World About Us: 34 right (Richard Taylor).

Illustrators Sandra Pond, Andrew Macdonald, Alan Suttie, Ann Savage. Paintings throughout pages 1–9 by Peter Morter. Maps by William Donohoe. New artwork: 24, 29 (Bridget Morley); 183 (Richard Bonson).

Temperature graph, 29: Source: Mann et al. (1999), Millennial Northern Hemisphere (NH).

Botanical plates Franz Antone, *Atlas Phytoiconographie bromeliaceen*, 32 top right. Vicente Martin de Argenta, *Album de la Flora medicofarmaceutica é industrial, indígena y exótica*, 18. R. Bentley and H. Trimen, *Medicinal Plants*, 84, 124. F. Cassone, *Flora medico farmaceutica*, 114. Crown copyright: reproduced with permission of the Controller of Her Majesty's Stationery Office and the Director of the Royal Botanic Gardens, Kew, 18, 46, 56, 58, 77, 97, 101, 122, 144. Curtis's Botanical Magazine, 32 top left, 65, 123, 137, 139, 141, 143, 145, 181. Joseph Descaisne, *Le jardin fruitier du Museum*, 82. M. Greshoff, *Nuttige indische planten* 119. Hedrick, *Plums of New York*, 64. Sir Joseph Hooker, 18. Nicholaus Jacquin, *Floral austriacae*, 18, 166. Fr. Eugen Kohler, *Medizinalpflanzen*, 91. Lindley and Paxton, *Flower Garden*, 50. Thomas Nuttall, *The North American Sylva*, 150. J. Parkinson, *A Garden of all sorts of Pleasant Flowers*, 136. Joseph Plenck, 46, 58, 82, 115, 139. Pierre Joseph Redoute, 90. François Regnault and Geneviève de Nangis Regnault, *La botanique mise à la portée de tout le monde*, 48, 64, 74, 84, 89, 90, 91, 105, 106, 113. John Sibthorpe, *Florae graecae*, 119, 140. J.T. Syme, *English Botany*, 83. Tanako Citrus Studies, 75. Ventenat, *Jardin de la Malmaison*, 181. William Woodville, *Medical Botany*, 138.